Culture, Curriculum, and Identity in Education

Culture, Curriculum, and Identity in Education

Edited by H. Richard Milner IV

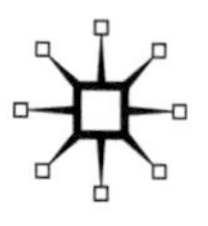

CULTURE, CURRICULUM, AND IDENTITY IN EDUCATION

First published in 2010 by PALGRAVE MACMILLAN® in the United States—a division of St. Martin's Press LLC, 175 Fifth Avenue, New York, NY 10010.

Where this book is distributed in the UK, Europe, and the rest of the world, this is by Palgrave Macmillan, a division of Macmillan Publishers Limited, registered in England, company number 785998, of Houndmills, Basingstoke, Hampshire RG21 6XS.

Palgrave Macmillan is the global academic imprint of the above companies and has companies and representatives throughout the world.

ISBN: 978-0-230-62204-3

Library of Congress Cataloging-in-Publication Data

Culture, curriculum, and identity in education / [edited by] H. Richard Milner.
p. cm.
Includes bibliographical references and index.
ISBN 978-0-230-62204-3 (hardback)
1. Multicultural education—Curricula. 2. Multiculturalism. 3. Critical pedagogy. I. Milner, H. Richard.

LC1099.C83 2010
370.117—dc22 2009041347

A catalogue record of the book is available from the British Library.

Design by Scribe Inc.

First edition: March 2010

10 9 8 7 6 5 4 3 2 1

Printed in the United States of America.

For the many lessons they taught me about culture and identity, this book is dedicated to the memory of my grandmothers, Eunice Milner, Annie Williams, and Corine Williams.

Contents

Illustrations

Foreword

Jacqueline Jordan Irvine, Emory University

I have spent my entire professional career grappling with issues related to educating teachers for the growing numbers of culturally diverse students in our nation's public schools. These culturally diverse students are African American, Latino, and American Indian students who rank at the bottom of most measures of school achievement. Data presented in this book point out that teachers who work in schools attended by students of color are mostly White females whose limited experience and preservice education programs have left them unprepared to teach in urban, high poverty, and diverse schools.

H. Richard Milner IV has compiled an outstanding volume, *Culture, Curriculum, and Identity in Education*, that analyzes equity and diversity in P through 12 schools and teacher education. Within this broad and necessary context, the book raises some critical issues not previously explored in many multicultural and urban education texts. These cutting-edge topics include, for example, gender, spirituality, English-language learners (ELL), bilingual education, immigration, and identity, as well as practical suggestions and implications for program implementation and revision, theory development, and research.

As I read the various chapters in the book, I reflected on the philosophy and writings of my mentor and dear friend, the late Dr. Asa G. Hilliard (Nana Baffour Amankwatia II). Dr. Hilliard was clear in his thinking that we could solve the "problems" of low achievement for students of African descent (and other students on the margins of learning and opportunity) by transforming ordinary solutions into extraordinary ones. He wrote, "It is clear that ordinary public school teachers, with unselected regular classrooms, serving poor children, without specialized standardized programs, can move students to the highest academic levels in a short period of time. It is not the children or their parents, poverty, culture or bilingual status (correlates that explain little or nothing) that determine academic success. It is good teaching" (Hilliard, 2000, p. 6).

Hilliard also understood that quality classroom instruction took place within the broader context of school and society. Referencing Kozol's *Savage Inequalities* (1991), he stressed, "Our children's manifest problems in public education virtually all have to do with opportunity to learn" (Hilliard, 2000), and "our highest research priority ought to be given to the study of what disables ordinarily effective pedagogy, disables systems that crush the spirits of teachers and the students, and blocks their natural genius" (Hilliard, 2007).

Hence, the challenge and the legacy that Hilliard left the P through 12 and teacher-education community were not only related to closing the achievement gap but also related to closing the other gaps that prevent culturally diverse students from experiencing school success. These gaps include the teacher-quality gap, the teacher-training gap, the challenging-curriculum gap, the school-funding gap, the digital-divide gap, the wealth and income gap, the employment-opportunity gap, the affordable-housing gap, the health care gap, the nutrition gap, the school-integration gap, and the quality child-care gap. He urged teacher educators to do the following:

- Define their role as social reconstructivists
- Teach their students how to dismantle systems of racism, inequality, and oppression
- Advocate for those who cannot advocate for themselves
- Tackle issues of structural inequality
- Address apathy, ignorance, and racism wherever they exist

These themes and the challenges embedded within them are emphasized in this book. Part I, "Identity and P through 12 Curriculum in Multiple Contexts," includes chapters by Toshalis (Chapter 1) and Milner (Chapter 2).

Toshalis's chapter explains how teachers react to the identity-perception gap. He raises the question: How do teachers describe a gap between "who I think I am" and "who students think I am"? The power of Toshalis's research is that his findings explore one of the most basic elements surrounding teacher-student interactions and their subsequent relationships. He explains that teachers' perceived identity impacts their ability to teach diverse students and their own sense of worth. Rather than avoiding the vulnerability and stress caused by this gap, Toshalis challenges us to take the necessary steps to view the gap as a resource for further engagement, exploration, and understanding.

Milner's chapter further explores the importance of teacher-student identity and interaction through his research on an African American teacher and the complexities of developing a multicultural curriculum in a predominantly White teaching context. The chapter provides important perspectives on how the teacher negotiates and balances her own beliefs and ideology in the curriculum in order to meet the needs of her mostly White, suburban students. Indeed, Milner's chapter challenges readers to consider the important interplay between teacher and student identity and what students have the opportunity to learn in school.

Part II, "Culture, Curriculum, and Identity with Implications for English-Language Learners and Immigration," contains three chapters. In Chapter 3, Iddlings and Rose report on an inexperienced elementary teacher in a research-based, professional-development project who attempts to work with a recent immigrant Mexican student enrolled in her class. The researchers conclude that some of the essential supports needed for the student's success include the unrestricted use of native language, a rigorous curriculum, a clear and explicit organization of the content, and opportunities for peer interaction. They point out that the challenges of implementing these supports are formidable given the current political climate focused on accountability and high-stakes testing.

In Chapter 4, Irizarry and Raible unpack the sociopolitical context of English-language learners (ELLs). Using the voices of students, their families, and elders, they shed light on the commitments and responsibilities that multicultural teacher educators must embrace in order to provide a meaningful education for ELLs. They push readers to face the more uncomfortable stances we take on issues related to ELLs and call for a teacher-education agenda based on "political clarity" about the sociopolitical context teachers and students encounter. This agenda would challenge marginalization and encourage teacher educators to be border crossers who develop strong ties with linguistically diverse students and their families. James, in Chapter 5, explores the experiences of two African Canadian college students and unveils contemporary issues seldom discussed in the literature. These issues include how culturally diverse immigrant students sometimes distance themselves from their community peers in order to succeed, and how they navigate and circumvent racism. James allows the participants in his study to reflect back on their schooling experiences in an urban context in order to gauge their struggles and success in college. Implications for immigration, identity, and culture are embedded throughout each of the chapters in this section.

The third part of the book, *Spirituality as Identity with Implications for Research and Teaching*, has two separate chapter authors: Dantley and Hancock. The chapter authors boldly consider an often-omitted feature of identity in the research literature—spirituality. For instance, in Chapter 6, Dantley calls into question the hegemony perpetuated by traditional qualitative research where issues of spirituality are concerned. Using the published works of Dillard, Tillman, Lather, and Milner, he constructs new ways to think about qualitative research that is positioned in an activist political agenda. Furthermore, he argues that spirituality is so entrenched in subjectivity and one's internal focus that traditional methods seem "self-defeating and pointless."

While Dantley focuses on spirituality in research, Hancock's focus in Chapter 7 is spirituality and teaching. He summons the reader to consider our educational system as morally bankrupt. He states that there are fundamental issues that teaching and learning should embrace—humanity, wholeness, and intellectual peace. Hancock brings to the fore these commonplace and accepted values present in noneducation settings but curiously absent in education. He proposes that the education community reflect on moral principles like respect, love, peace, and care. Indeed, he believes that critical reflection is a spiritual process that helps teachers support students' personal and academic success. Additionally, critical reflection assists teachers in adopting social activism as part of their professional responsibility.

The final part of the book is titled *Culture, Curriculum, and Identity with Implications for Teacher Education.* This final part has chapters by Milner (Chapter 8) and Cross (Chapter 9). Milner considers race as an identity marker in teacher education and walks readers through a process of inquiry where he attempts to study his own practice, particularly his curriculum development and implementation, in order to improve it. The chapter challenges teacher educators to consider their own identity in their work, to address the salience and presence of race as identity in teacher education, and to reenvision the work of the curriculum in their courses and in the teacher education program. In the final chapter of the book, Chapter 9, Cross explores the meanings of the racial, ethnic, and cultural mismatch between the mostly White teaching force and the growing, diverse P through 12 student body. Using the often-ignored voices of African American, Latina, and Native American teachers, she discovers that the teachers are in agreement that extraordinary efforts are required to reframe teacher education from its current focus on human relations to promote diversity to a bolder and more reformative agenda that prepares teachers with the efficacy and agency necessary for systemic and structural change.

The astute analyses and compelling recommendations proposed in this book would resonate with Asa Hilliard. The authors do not focus on simplistic strategies and programs. Instead, they present bold proposals rooted in social justice and excellence that require commitment and courage by individuals. Hilliard reminds us that excellence comes from people and not programs and advises us to forget about decoys, robot-like approaches, and "one-trick ponies" like high-stakes testing, vouchers, charters, commercial programs, remedial work, and individual educational programs (Hilliard, 2003, 2007).

Finally, the book underscores the courageous vision necessary for excellence and equity, a vision eloquently described by Hilliard (2003): "If we love the children, then we must do whatever it takes to provide them with the teachers and school leaders they deserve. We cannot tolerate or support ideologies and practices that cripple our children further—those that hold our children as the problem or those that assume that our teachers and school leaders are not capable of becoming powerful factors in the lives of students. We need a valid vision. We need the will. With vision and will, everything is possible" (p. 165).

Asa Hilliard's philosophy and vision are expressed throughout the chapters in this volume, and readers are presented with clear and explicit plans of action for social justice and equity for all children. Indeed, *Culture, Curriculum, and Identity in Education* provides readers with research-based approaches that are grounded in theory and that can make a real difference in the lives of teachers, teacher educators, and students.

References

Hilliard, A. G. (2000, April). *The state of African education.* Paper presented at the meeting of the American Education Research Association. New Orleans, LA.

Hilliard, A. G. (2003). No mystery: Closing the achievement gap between Africans and excellence. In T. Perry, C. Steele, and A. G. Hilliard (Eds.), *Young, gifted, and Black: Promoting high achievement among African American students* (pp. 131–165). Boston: Beacon Press.

Hilliard, A. G. (2007, April). *Shaping research for global African educational excellence: It is now or never.* Paper presented at the meeting of the American Education Research Association, Chicago, IL.

Kozol, J. (1991). *Savage inequalities: Children in America's schools.* New York: Crown Publishers.

Acknowledgments

I am forever grateful to my wife, Shelley Banks Milner; my parents, Henry III and Barbara; and my sister, Tanya Milner McCall. Your true acts of love and support make it possible and necessary for me to do this work.

I am also grateful to my major professors during my doctoral studies at The Ohio State University: Dr. Gail McCutcheon and Dr. Anita Woolfolk Hoy. I am also grateful to a true mentor, Dr. Sarah Favors, from South Carolina State University. Your unwavering support of my work makes the difference. Thank you!

Introduction

Culture, Curriculum, and Identity in Education

H. Richard Milner IV, Vanderbilt University

Culture

This book is about the intersections of culture, curriculum, and identity in education. Individuals and groups of people operate in and through cultural frames of reference and in and through social contexts. Our beliefs, ideologies, ways of knowing, preferences, and practices are shaped and guided by culture. For reasons elaborated elsewhere (see Chapter 8 for instance) but beyond the scope of this introduction, White teachers sometimes do not believe and fully understand that they have a culture (see, for instance, Milner & Smithey, 2003), or that their worldview and practices are culturally grounded, guided, and facilitated. They struggle to understand that they, like people of color, too are cultural beings and that their conceptions, decisions, and actions are culturally shaped and mediated. They sometimes classify others as "cultural beings" or "diverse" and sometimes do not recognize the salience and centrality of their own culture, and how it is woven through their work as teachers. Culture is steeply embedded within and around each of us, is in and among all groups of people, and is especially shaped by the social context of education.

Culture is not a static concept—"a category for conveniently sorting people according to expected values, beliefs, and behaviors" (Dyson & Genishi, 1994, p. 3). Rather, culture is ever changing and ever evolving, and, fundamentally, the contributors of this book attempt to convey the message that the work of teachers, teacher educators, researchers, students, parents, and principals is deeply guided by culture and contexts. We are cultural beings and workers. Accordingly, we must engage our work in ways that allow us, as Irvine (2003) asserted, to see (and live) with a cultural eye. Culture encompasses various other concepts that relate to its central meaning. For instance, culture may include individuals or a range of a group's identity markers such as class, socioeconomic status, gender, race, sexual orientation, and language. Building on and from the work of a number of other researchers and theorists (see, for instance, Dyson & Genishi, 1994; Irvine, 2003; Nieto, 1994), I conceptualize culture as the implicit and explicit characteristics of a person or group of people—characteristics developed through historic, sociocultural backgrounds, current experiences, knowledge, disposition, skills, and ways of understanding. These characteristics and ways of being are informed by race, ethnicity, history, heritage, customs, rituals, values, symbols, language, identity, class, region/geography, resources, and gender.

It is important to note that I do not mean to imply that culture is synonymous with race, although race influences people and group's cultural experiences. However, culture is much more dynamic in terms of the multiplicity of guiding factors. Still, race is an important dimension of culture. Because they do not necessarily see themselves as cultural beings and because they do not believe they are governed by a culture, teachers may see themselves as "the norm" (Foster, 1999). Consequently, their curriculum development and curriculum enactment are designed to have their students work to catch up to that norm, which are ironically their cultural frames of Whiteness. Similarly, even though teachers of color typically see themselves in light of culture, they sometimes fail to recognize or to understand how culture shapes their practices with their students. They understand, for instance, that culture influences their life experiences outside the classroom but sometimes fail to transfer those conceptions in their actual teaching practices with students. Thus, this book treats culture and the related themes explored herein as relevant to all educators, not just a specific group of them.

Curriculum

Not only is this book about culture, but it is also about the curriculum (for a much more detailed discussion about the nature of the curriculum and its implementation, read Chapters 2, 3, and 8). By curriculum, I mean *what* students have the opportunity to learn in educational institutions and in related contexts across the United States and abroad. Consider, for instance, Eisner's (1994) postulation of several forms of the curriculum: (a) the explicit curriculum concerns student-learning opportunities that are overtly taught and stated or printed in documents, policies, guidelines, and Web sites; (b) the implicit curriculum is intended or unintended but is not stated or written down and can also be considered the hidden curriculum, which occurs through learning opportunities that are implicit and covert in a context; and (c) a third form of curriculum, the null curriculum, deals with what students do not have the opportunity to learn. For me, this form of Eisner's curriculum, the null curriculum, is the most powerful. Information and knowledge that are not available for student learning is also a form of the curriculum. What students do not experience becomes messages, information, and data points for the students. For example, even if students are not being taught to question power structures, the students are still being taught about questioning power structures—possibly that it may not be essential to question or to critique them. Stories told (or not), then, are a critical part of both the implicit and null curriculums—they influence learning opportunities for students. *What is absent or not included is actually present in what students are learning.*

To elaborate, if adults never have serious and tough conversations with students about anti-Semitic comments, the students are learning something—perhaps either that such a topic is not important or that it is inappropriate to consider or discuss that topic. When students are not taught different points of view in the curriculum, they are learning something as well. The ways in which people interpret silence can vary. For example, when adults witness racist, insensitive comments and say nothing, neither formally nor informally, they are in essence communicating and teaching their students something. Delpit (1995) provided a powerful perspective that extends this point of view in what she called the silent dialogue. People who make the insensitive comments are communicated with even though the silent voices can have multiple meanings and intentions. Those who communicate the racist and insensitive language may believe that silence is a sign of acceptability; they may believe, based on silence, that their language is tolerable because no one has spoken out against it. Freire (1998) asserted that it is the *responsibility* of those who have a voice and consequently power to speak out and to use it when they witness and/or experience injustice. Moreover, students should be empowered to use their own voices and to speak out against situations they believe are inappropriate. Talk is behavior and action, and adults should be challenged and have the conviction to take action when they witness injustice. The idea of *speaking up and speaking out* is especially important when the victim, the person on the receiving end of the hurtful language, cannot speak out for herself or himself. To recap, *silent voices are voices that are speaking—sometimes loudly* and can be considered part of the null curriculum.

I will try to position readers to transfer and transcend school/classroom-related understandings of curriculum with that of other non–school related contexts such as the learning that takes place in nonprofit organizations, summer camps, on the fields and courts of sporting events, daycare centers, and adult-oriented environments. The educational institutions that are represented in this book include P through 12 schools, those areas outside of school educational contexts, as well as higher education (namely teacher education). Again, the curriculum has to do with students' learning opportunities and also with their access to opportunities to learn. Learning opportunities emerge in different and dynamic, vibrant and ever-changing spaces. Students learn from their parents in their homes (and thus a curriculum is enacted in students' homes). Students learn from their peers in the corridors and common spaces in school (and thus a curriculum is enacted in the corridors and common spaces) as well as on the basketball court and lacrosse field during recreational activities (and thus a curriculum is enacted on the court and on the field). Thus, learning opportunities are not exclusively available to students in a classroom with the teacher "giving" information through some organized curriculum guide. Learning opportunities are available at different times and in different spaces.

In school, teachers, too, are learners who are mediated by culture and identity. Teachers rely on their cultural references when they teach—their pedagogical approach is shaped culturally by how teachers learned as students, how they teach and convey information to their own biological children, and what they emphasize in the curriculum over other information. Moreover, as readers will come to understand, students and teachers themselves are texts (written by their historical and current life experiences), and students have the opportunity to read and to learn from the texts of their teachers and classmates. In this sense, *teachers and students are a form of the curriculum themselves*, which makes concentrating on teachers and students' cultures and identities profoundly important. This point can easily be seen and understood when researchers stress the need for African American and other students of color to interact with and to be taught by successful African Americans and other teachers of color. The argument is that these students need to "see," witness, and experience these other teachers of color—to read their texts—because the teachers' experiences allow students to learn something important based on their interactions with them. Thus, teachers (similar to other individuals) are essentially cultural beings who are curricula texts: students have the opportunity to learn something based on their interactions with and readings of these teachers and vice versa.

Identity

Culture and identity are akin. While culture can be seen as a concept that categorizes or describes traits or characteristics of a person, it usually is also a concept that captures a group of people. Identity is a concept that is often used to classify individuals, although group identities are also categorized and critiqued at times. Identities have multiple features and markers to them, and individuals often operate through and are guided by various (cultural) identities. Identities, similar to cultures, are always evolving and emerging: people are continuously developing their identities based on their experiences; based on what they come to know about themselves, others and the world; and also based on their developmental ideologies and worldviews. For instance, people are consistently developing their cultural identity, their language identity, their spiritual identity, their racial and ethnic identity, their professional identity, and their gender identity both implicitly and explicitly. Social contexts, of course, help shape the evolution and development of identity.

Therefore, teachers are always developing their teaching identities, and students are constantly developing their identities as students and learners. Teachers decide what kinds of teachers they want to be and the kinds of educators they want others to perceive them as being (read Chapter 1 for more on this important issue). Teachers decide how they want to be perceived by their peers, the parents of their students, their students, and also their administrators. Teachers also "perform" for researchers. They try to live up to an identity they think they want to portray to observers. Teachers and students, for instance, define and view themselves in particular ways and subsequently represent themselves to others in ways that carry varying interpretations. Student identities encompass their social identities. Students decide whom they are going to "hang out" with, whom they want to date, what social events they will attend, and whom they will invite to those social events. In some sense, identities are "acted out" based on whom people decide they want others to get to know. Identities can even be ascribed to people, and individuals can either accept or reject the ascribing. Thus, this book attempts to disentangle some of the complexities inherent in both teachers and students' conceptions and representations of their multiple and varied identities and what these identities mean in, through, and for the learning opportunities available in a particular context.

Researchers study and conceptualize identity in a range of important ways depending mostly on their epistemological positions—the way they are positioned to see, experience, understand, and represent their realities to others. For example, from a psychological perspective, identity is viewed through lenses of mental imaging and physical characteristics of an individual. Researchers in this line of thinking focus on "self" characteristics such as how people define and construct themselves. Sometimes absent from this line of thinking regarding identity are social forces that arbitrate the self-identity. However, social cognitive theories consider how social influences shape self-thinking regarding identity. Sociological perspectives view identity through lenses of the social: society and contexts inform and produce identity. Identity, for some sociologists, is married to society, and the interplay between society and identity are the concentration of this research and thinking. Thus, as readers engage this book, they will see that identity can be conceived of in different ways.

The authors of this book understand the interplay between culture, curriculum, and identity. Their contributions draw from empirical research to frame important recommendations for ways to think about and to address the many issues embedded in and through matters of culture, curriculum, and identity in education.

This Book and the Audience

Renowned national and international scholars, Jacqueline Jordan Irvine provides an introduction to each of the chapters in her foreword, while Sonia Nieto provides the afterword. In this introduction, I briefly discuss each section of the book and outline framing questions. For a more in-depth look at each chapter, read Irvine's foreword, and for a well-grounded reflection of the book, read Nieto's afterword.

The book is divided into four interrelated parts. Part I, "Identity and P–12 Curriculum in Multiple Contexts," has two chapters and focuses on matters of identity in learning opportunities and the various contexts that shape identity development and enactment. Central questions that guide this section include the following:

- How do teachers negotiate and balance who they think they are and how students perceive them in the learning context?
- How do teachers and students' identities shape learning opportunities in the classroom?
- How do teachers negotiate and balance their own cultural identities in their decision making?
- What are the implications of teachers' varied and multiple cultural frames for various learning contexts?

Part II, "Culture, Curriculum, and Identity with Implications for English Language Learners and Immigration," has three chapters. This part focuses on bilingual education, English-language learning and learners, and immigration. Clearly, the population of the United States is becoming increasingly diverse, and with this diversity comes a rich and important range of language and identity among teachers and students alike. The inextricably linked issues of language and identity must be considered as the demography and immigration patterns of the United States changes and intensifies. Banks (2006) explained that "the source of the nation's immigrants has changed substantially. Most of the 8.8 million immigrants who came to the United States between 1901 and 1910 were Europeans . . . Between 1991 and 2000, 82 percent of the documented immigrants to the United States came from nations of Asia, Latin America, the Caribbean, and Africa . . . The U.S. Census Bureau projects that ethnic groups of color will increase from 28 percent of the nation's population today to 50 percent in 2050" (p. xvii).

Central questions that guide this part include the following:

- Why is it so important for students to become bilingual and not focus on English-language learning to the exclusion of students' native language?
- What historical, political, and sociocultural landscapes shape language learning in U.S. society and classrooms?
- How do teachers develop teaching strategies that honor students' native language and, at the same time, provide students' opportunities to learn English in order to function in (and transform) U.S. society?
- What can teachers do to maximize students' language-learning opportunities and also maintain and honor students' cultural identities?

Part II also considers the schooling experiences of immigrant Black students from an urban neighborhood. Immigration matters and language dilemmas are the central themes of this part. Readers are invited to explore some complexities inherent in the negotiations in which both students and teachers must engage to maximize access and learning opportunities for all students.

With two chapters, Part III, "Spirituality as Identity with Implications for Research and Teaching," concentrates on the absence of spirituality in much of the literature on culture and suggests that spirituality be considered in both research studies and in teaching. In the academy, a lack of understanding about how to study the spirit or a lack of consensus on what the spirit is or whether it exists is not sufficient for ignoring and un(der) exploring spirituality in our work, particularly when discussing culture and identity in research, education, and teaching. Spirituality has served as a source of resistance and survival for some groups of people throughout history. For example, spirituality has been a source of survival for many African Americans in the United States and abroad, and this should not be ignored when studying their experiences and life worlds (Dantley, 2003; Dillard, 2002). Central questions that guide the chapters in this part include the following:

- What role might spirituality play in research that attempts to answer salient and complex questions regarding identity, equity, and social justice in education?
- How is spirituality a form of identity in the work of researchers and teachers?
- What are the relationships between identity and spirituality in our work as teachers, teacher educators, and researchers?
- How might spiritual positioning be used as an asset in schools that work to meet the needs of all learners?

The final part of the book, Part IV, "Culture, Curriculum, and Identity with Implications for Teacher Education," considers matters of culture, curriculum, and identity for students of teacher education and teacher educators themselves. It has become apparent that teachers and students of teacher education need to be better prepared to teach and meet the needs of their respective students. An impetus for this charge and need can be meaningfully linked to the demographic divide that exists between teachers and students in P through 12 classrooms. The demographic divide is present in an important body of literature that makes a case for the preparation of teachers for the diversity they will face in P through 12 educational contexts (see, for instance, Zumwalt & Craig, 2005).

These demographic divide data include gender, race, ethnicity, and socioeconomic background. For the purposes of this discussion, Tables I.1 and I.2 provide racial demographic data of public school teachers and students. In short, although the demographic divide is much more complex, space in this introduction allows me to provide as an example an emphasis on the increasing number of White teachers and non-White students. Central questions that guide the chapters in this section include the following:

- How do teacher educators study their own practices to provide a curriculum that enables their students to understand how identity markers can infuse and transform the P through 12 curriculum?

- What are some lessons embedded in the identities of teachers of color that can be instructive for other teachers, particularly those teachers teaching in urban and highly diverse learning environments?
- How can teacher-education programs recruit a more diverse teaching force and, at the same time, meet the complex needs of their mostly White teachers who may find themselves in teaching contexts with highly diverse students?

Table I.1. Teaching demographics in public elementary and secondary school, 2003–2004

	Elementary public school (%)	*Secondary public school (%)*
White	81.6	84.2
Black	8.8	7.5
Hispanic	7.0	5.5
Asian	1.3	1.3
Pacific Islander	0.2	0.2
American Indian/Alaskan Native	0.4	0.6
More than one race	0.7	0.7

Note. Institute of Education Sciences: National Center for Education Statistics. (n.d.). *The condition of education.* Retrieved December 6, 2007, from http://nces.ed.gov/programs/coe/2007/section4/table.asp?tableID=721

Table I.2. Student demographics, 2003–2005

	2003 (%)	*2004 (%)*	*2005 (%)*
White	60.5	59.9	59.4
Black	14.9	14.9	14.8
Hispanic	17.7	18.2	18.7
Asian	3.6	3.7	3.7
Pacific Islander	0.2	0.2	0.2
American Indian/Alaskan Native	0.9	0.9	0.9
More than one race	2.2	2.3	2.3

Note. Institute of Education Sciences: National Center for Education Statistics. (n.d.). *Digest of education statistics: 2006.* Retrieved December 6, 2007, from http://nces.ed.gov/pubsearch/pubsinfo.asp?pubid=2007017

It is impossible to completely answer the questions posed in this introduction—questions that frame the different parts of this book. However, the posing of these questions and our attempt to begin to respond to them matter more than readers finishing the volume with complete answers. I am hopeful and optimistic that readers begin to address these questions themselves based on their own experiences, positioning, worldviews, cultures, and identities. And I am also hopeful that readers will develop their own more contextualized, nuanced, and elaborated questions and become more intimately aware of the interactions between and among culture, curriculum, and identity. Indeed, cultural, identity, and curricula work is lifelong work that will require continuous reflection on these and other questions!

The question remains: Who should read this book? Teachers, teacher educators, researchers, and other educators interested in the interplay of culture, curriculum, and identity in education will find this book insightful. The authors draw from research studies to build theory, critique reality, and provide practical, real-life recommendations for readers. P through 12 teachers will find that chapters from each section speak to current school matters with and about a range of students in different spaces across the United States and abroad. Teacher educators will find that readers are invited to visualize what *can be* in schools, and how teacher educators can serve as leaders in the fight for social justice–oriented curriculum development and implementation. Researchers are challenged to pose different kinds of questions—questions that look at the possibilities rather than those of despair and hopelessness in their work to address, redress, and transform institutional and systemic inequality, inequity, oppression, marginalization, and discrimination.

Taken together, the chapters in this book are written, from my perspective, by some of the most influential, promising, and committed teachers, teacher educators, and researchers in the United States and abroad. I invited individuals to contribute to the book whom I know care deeply about the state of education for *all* students—scholars who understand the relevance and centrality of culture, curriculum, and identity in education. These scholars consider the multiple, varied, and complex issues and needs that students and teachers face both inside and outside of the classroom. They take very seriously their individual roles and understand their collective responsibility to speak truth to oppressive and hegemonic powers that can prevent both the student and the teacher from reaching their full capacity.

References

Banks, J. A. (2006). *Cultural diversity and education: Foundations, curriculum, and teaching* (5th ed.). Boston: Pearson Education.

Dantley, M. E. (2003). Critical spirituality: Enhancing transformative leadership through critical theory and African American prophetic spirituality. *International Journal of Leadership in Education, 6*(1), 3–17.

Delpit, L. (1995). *Other people's children: Cultural conflict in the classroom.* New York: The New Press.

Dillard, C. B. (2002). Walking ourselves back home: The education of teachers with/in the world. *Journal of Teacher Education, 53*(5), 383–392.

Dyson, A. H., & Genishi, C. (1994). *The need for story: Cultural diversity in classroom and community.* Boston: Harvard University Press.

Eisner, E. W. (1994). *The educational imagination: On the design and evaluation of school programs.* New York: MacMillan College Publishing Company.

Foster, M. (1999). Race, class, and gender in education research: Surveying the political terrain. *Educational Policy, 13*(1/2), 77–85.

Freire, P. (1998). *Pedagogy of the oppressed.* New York: Continuum.

Irvine, J. J. (2003). *Educating teachers for diversity: Seeing with a cultural eye.* New York: Teachers College Press.

Milner, H. R., & Smithey, M. (2003). How teacher educators created a course curriculum to challenge and enhance preservice teachers' thinking and experience with diversity. *Teaching Education, 14*(3), 293–305.

Nieto, S. (1994). Lessons from students on creating a chance to dream. *Harvard Educational Review, 64*(4), 392–426.

Zumwalt, K., & Craig, E. (2005). Teachers' characteristics: Research on the demographic profile. In M. C. Smith & K. M. Zeichner (Eds.), *Studying teacher education: The report of the AERA panel on research and teacher education* (pp. 111–156). Mahwah, NJ: Lawrence Erlbaum.

Part I

Identity and P through 12 Curriculum in Multiple Contexts

1

The Identity-Perception Gap

Teachers Confronting the Difference between Who They (Think They) Are and How They Are Perceived by Students

Eric Toshalis, California State University Channel Islands

> To face the provisionality and contingency of identity is to surrender the security of stable identities, identities that are often necessary if one is not to be worn down by the daily grind of living in a racist society.
>
> —P. Taubman, "Facing the Terror Within: Exploring the Personal in Multicultural Education"

Urban teaching is identity work. Everyday, teachers confront social, psychological, and political challenges as they try to promote the achievement of hundreds of students who emerge from a multiplicity of backgrounds. To achieve equity in schools and classrooms requires that teachers sustain a near-constant awareness of others' unique perceptions and identities. As teachers negotiate classroom relationships with youth across cultural, gender, linguistic, racial, sexual, and socioeconomic differences (to list only a few), identities get exposed, deconstructed, and disrupted in the classroom as a matter of routine. Although these differences may produce profound breakthroughs in understanding and practice, they can also generate tremendous anxiety in those facing questions about who they are, who they should be, and how they want others to see them.

Often, formative questions about identity are thought to be the developmental hallmarks of adolescence, immersed as many teens sometimes are in

The author wishes to thank Wendy Luttrell and Janie V. Ward for their help in developing the ideas in this chapter and editing early versions of this work.

"identity crises" (Erikson, 1968), but the truth is that teachers struggle with such issues all the time. For example, what does a teacher do when the identity she understands herself to possess is not affirmed in her relationships with students? What happens to those relationships when there is a gap between who teachers think they are and how others perceive them? How should they make sense of this "identity-perception gap"? The ways teachers answer such questions, what they feel their relationships with students say about who they are, and which practices they adopt to maintain those relationships are critical to establishing equity and valuing diversity in the classroom. For this reason, a close analysis of this identity-perception gap is warranted.

In this chapter, I will explore the identity-perception gap, how teachers understand and react to it, and how it threatens their sense of self and self-worth as teachers of urban students. To do so, I will analyze a series of teacher conversations recorded by the research team of Project ASSERT (Accessing Strengths and Supporting Effective Resistance in Teaching) at the Harvard Graduate School of Education. The principal investigators of Project ASSERT, doctors Wendy Luttrell and Janie V. Ward, designed the five-year study to gather and examine teacher talk at several Boston-area public high schools by sponsoring conversations among teachers who were concerned about issues of equity, diversity, and identity in the classroom. Sometimes organized around researchers' suggestion for a theme or activity, and other times open to the participants' curiosities and needs, the conversations typically lasted ninety minutes and occurred after school approximately once per month during 2002 and 2003. In 2003 and 2004, the ASSERT team transcribed all meetings, created a coding manual, and established interrater reliability. "Interrater reliability" is a term researchers use to evaluate and enhance the level of homogeneity among different observers as they analyze the same phenomena. In Project ASSERT's case, the research team met to compare how each individual researcher coded a series of the same texts until we established agreement on how such passages should be labeled. We also instituted a refinement process whereby we would bring nebulous passages to the collective for group analysis.

Research members such as myself were then encouraged to explore themes in the data that interested them. I chose to analyze the conversations of one particular group due to the diversity of its membership, the school's urban location and student population, and my familiarity with the dynamics of the group (I had attended several of their after school meetings as a note taker in the second year of data-gathering).

During my analysis of fourteen conversations recorded over a two-year period, I noted the prevalence of what I am calling the *identity-perception gap* in the teachers' descriptions. This first stage of analysis led me to return to the data with this question: *How do these teachers describe a gap between "who I think I am" and "who students think I am," and what happens to their conversation when such a gap is identified*? My analysis suggests the presence of two main themes that I will explore in this chapter: the pervasiveness of vulnerability as a way of characterizing the identity-perception gap and a form of "comfort" the teachers provide for one another that inhibits deeper conversations about their own and others' different perceptions.

Theoretical Background: Identities and Perceptions in the Classroom

In my experience as a teacher, researcher, and teacher educator, I have found that when teachers perceive an affirmation of their identity in their interactions with students, smoother communication, closer connections, elevated learning, and a general reduction in classroom stress results. However, when teachers encounter a gap between "who I think I am" and "who students think I am" they often face anxiety-producing questions that interfere with the potential for learning, both the teacher's and the students'. Faced with this identity-perception gap, teachers may ask the following questions:

- Should I accept the validity of my students' perceptions of me, or should I reject it?
- If I choose to reject their perception of me, on what grounds do I base that rejection?
- If I choose to acknowledge my students' perception of me, how do I make sense of the gap between their perception and my perception?
- What might this gap suggest about me and my teaching?

Questions like these reveal how teachers confronting the identity-perception gap are often pushed to the limits of their understanding of identity itself. The modern, Western construction of identity as a singular, cohesive, durable "me" across multiple domains sometimes loses its explanatory power when teachers confront the ways their students interpret, label, and evaluate them differently than they do themselves. This may be why teachers so often express the need for what is suggested in the epigram—secure and stable identities that can withstand myriad interpersonal interactions that occur in classrooms on a daily basis, some of which may be prejudicial.

Many of those working as teachers or with teachers can testify to the enormous five-days-a-week effort needed to invest in relationships with students and to create unique learning opportunities for audiences that are sometimes quite critical. When the teacher's investment is rejected by students for whatever reason, it may lead to interactions charged with power struggles and/or tinged with resentment. In middle- and high-school settings in particular, classroom interactions occur with youth who are attempting to discern where the boundaries of acceptable behavior are, who are immersed in the search for an identity they can call their own, and who are emerging into a consciousness that is able to entertain possibility, multiplicity, and new levels of complexity. Thus, teachers face challenges not just to their pedagogy or curricular choices, but to their worldview and self-view at every turn.

Try as they might, it is no surprise that many teachers engaged in the identity work of teaching find it difficult to maintain their "game face." The often unspoken expectation that teachers sustain a stoic consistency in their affect and composure can be difficult to realize when classroom events cause feelings to surge. Though a teacher's day may include numerous moments of delight and humor, emotions such as rage, frustration, and despair may be as common in many teachers' experiences as tardy bells, absent slips, and intercom announcements. Difficult emotions like these are often drawn to the surface when teachers feel as though their identities have become threatened, destabilized, or dismissed in their interactions with students. This may be because teachers sometimes estimate their success in the classroom according to how small the gap is between what is transmitted and what is accepted. If the teacher can reduce or eliminate gaps in perception, thereby equalizing the teacher's and the students' understanding of a concept, the teacher is believed to have succeeded. For example, students are frequently understood to "pass a test" when they reflect back to the teacher the content as it was delivered. This same relationship often occurs with regard to the teacher's identity. The teacher experiences a sort of safety in congruency when there is no difference between who she thinks she is and who the students perceive her to be. "I am an inspiring, hard-working, caring educator," the teacher might think to herself, "and my students know this." Were the students to demonstrate to that teacher that they perceive her in a less-flattering light, the teacher faces hard questions about how to make sense of that gap in perception.

As an overcompensation for the vulnerability experienced in classroom relationships where critical concerns regarding truth and identity might become negotiable, new and veteran teachers alike often display a relentless fixation on lesson planning, assessment, and classroom management. Sometimes, the need to control trumps the need to connect. In such cases, gaps in perception may be framed as impediments to rather than opportunities for learning and understanding. Teachers with this orientation may reason that if a student misunderstands a concept, it is the teacher's job to correct it; likewise, if a student misunderstands who the teacher is, it is the teacher's job is to correct that misinterpretation. It is believed there are right answers to the questions "What is to be learned?" and "How am I to be understood?" and it is the teacher who possesses the sole authority to determine them.

The trouble with this orientation toward identity and knowledge, according to Elizabeth Ellsworth (1997), is that it fails to take into account the extent to which our perceptions "constantly and inevitably [pass] through the uncontrollable stuff of desire, fear, horror, pleasure, power, anxiety, fantasy, and the unthinkable . . . [and this is] *exactly* what most educators stay up late on school nights trying to plan *out* of the next day. Classroom acts and moments of desire, fear, horror, pleasure, power, and unintelligibility are *exactly* what most educators sweat over trying to prevent, foreclose, deny, ignore, close down. Such stuff is scary to teachers" (p. 46).

To orient oneself to the classroom primarily as a teacher of content (as opposed to a teacher of people) is an approach that minimizes the potential for interpersonal vulnerability, whereas entering that space intent on building relationships with students necessarily involves some social and psychological risk. These two orientations, rather than being polar opposites, are simultaneously essential to the sorts of classroom practices that aim to achieve equity and enhance diversity. Fully engaging the relational side, however, requires us to acknowledge that teachers as well as students experience vulnerability in the classroom. It is to that subject that I now turn.

Vulnerability: Discomfort and Confusion in Teaching Across Difference

When confronting the gap between "who I think I am" and "who students think I am," teachers in this study often describe feeling vulnerable. Expressed as insecurity, exposure, or a susceptibility to misrepresentation, dismissal, or ridicule, these feelings of vulnerability seem to provoke in the teachers a heightened sensitivity to how their and others' perceptions are shaped by cultural, gender, racial, sexual, and socioeconomic categories. In the teachers' conversations, this awareness sometimes leads to insights about their relationships with students, but it also produces a sort of paralysis in their thinking such that they cannot conceive of what to say or do next. Faced with a student's condemnatory perception of who they are, teachers sometimes get stuck.

Kristina (this is a pseudonym, as are all participants' names in this chapter) is one of those teachers. A tenth-grade math teacher and out lesbian of German heritage who is well regarded by her peers, she considers herself "one of the best math teachers [at the school]." "Out" is used here to denote the fact that Kristina decided long ago not to hide her sexuality from her students and fellow faculty. It is well known at her school that she is gay.

Proud of her ability to promote academic success, she remains concerned about the underachievement of African American and Latino boys in her classroom. Lamenting the fact that she's "heard from several kids of color, 'Oh, yeah, you know, that those are the smart kids. The White kids are the smart kids. The White kids can do the math,'" Kristina is intent on making her classroom a place where math achievement is valued more than resisted. She recognizes, however, that mathematics carries racialized and gendered valences, stating, "In math there's a stereotype anyway, you know? Math is not for girls. Math is not for kids of color, you know? And I feel like one of my things is that I'm working hard to make kids believe they can do math, you know? And now I have this issue that is really, it's—I feel like in all my years at [this school], I've never, [this resistance has] never been so extreme." First framing her students' resistance as a matter of math's perceived racialized and gendered applicability, Kristina eventually divulges that it may be more personal than that. In the following excerpt, Kristina describes her feelings when she confronts what she perceives to be the students' perception not of math, but of *her*:

> So I'm sitting there working with these two girls and I tried to get the guys who were sitting at the same table to do work and they wouldn't. And I said several times, "John, can you get to work now? Can you get to work now?" And I just really felt like, you know, they were having a negative resistance because their conversation about basketball just got louder and louder . . . But I felt myself, this White woman nagging these young men of color to do math work, while they were talking about basketball, you know? And I just didn't know, you know, how to overcome that at that moment, you know? And I was sort of, I was projecting. I was projecting that they're thinking now, "Oh, there's this White lady that I don't even know from X coming here trying to tell us to do work." And it put me into this strange situation where I really didn't know how far to push and I became insecure.

Kristina demonstrates that she is sensitive to the racialized and gendered possibilities in how she may be perceived by her students, but that awareness seems to frustrate and confuse her. Believing that her students see her primarily as a "White woman nagging these young men of color to do math work," she confronts the possibility that no matter what actions she chooses next, her students may lay a different claim to her identity, possess a negative evaluation of their relationship, and form a much less forgiving appraisal of her motivations and/or character. Her ability to act with confidence seems to be impeded by her assumption that the boys have reduced their perception of her identity to race and gender.

Wanting simply to convince the boys to do their work and perform well in math, Kristina is suddenly made aware of the prospect that she is not viewed as the committed math teacher she understands herself to be. Her race and gender are foregrounded such that her identity as a talented math teacher seems almost irrelevant. The effect of her students' disengagement on their futures seems to be less important in that moment than her concern about their perception of her. This shift in a teacher's attention, even if it is only fleeting and especially if it goes unacknowledged and unanalyzed, can divert precious pedagogical energies toward ego preservation and away from safeguarding the academic success of a teacher's more-resistant students. As evidence of this diversion, Kristina admits that her understanding of the identity-perception gap may indeed be "projection," but this acknowledgement is still leaving her with no clue as to what to do. Regardless of her intent, she faces the fact that she is not in control of how her identity is perceived, and this leaves her vulnerable to the ways her students may (mis)represent and resist her despite the fact that she believes her teaching benefits them.

In subsequent conversations, Kristina adds that her attempts to get the students to do their work are frequently met with a resistance that has more to do with her perceived identity than with the content she teaches. In those moments, it is as if her classroom suddenly becomes a place less about math learning and more about identity negotiation, and this makes her uncomfortable (for insights into the negotiation of identities in classroom settings, see Alsup, 2006; Bernstein, 2000; Bingham, 2001; Dance, 2002; Ellsworth, 1997; Ferguson, 2000). She explains, "I'm a White woman and I'm telling a kid of color something to do. They may be resisting that because I'm a White person, and they don't want to hear from a White person." Resonating with such experiences and the emotions that arise in them, Shannon, a veteran English teacher who identifies as Irish, says, "The kids have their perspectives and their emotions, but the teacher has them too. And, sometimes, I just get pissed off."

Trying to translate their emotional reactions into some sort of pedagogical response, Kristina and Shannon are unsure about how much attention to pay to these gaps in perception. Do they matter? Should they matter? Kristina remarks, "It's really hard to have the radar that can really detect all the time. You know, well, what's the right thing I say at this point? 'Cause you're not really sure of what's going on." Concurring with this sentiment during another discussion among the participants, Shannon admits that in order to deal with the "complexity of these conversations [with students] . . . You have to deal with all of the complexities inside of you and, then, everything that they're putting on top of that. And I just think that's really, really challenging." Confused by the identity-perception gap, vulnerable to the students' appraisals, and aware of rising emotions that may not prompt their best pedagogical responses, Kristina and Shannon sometimes get stuck. When asked later how she understands her students' apparent resistance, Kristina responds, "It's boggled my mind. I mean, just because I have been confronted with it . . . As a White educator, I'm like 'Well, hmm!' I mean, I'm reinforcing the concept of White education by my appearance. Now, that doesn't mean my actions can't say something else, but yet my appearance immediately, you know, speaks for itself." Having expressed a desire not to teach in a way that valorizes White, mainstream, hegemonic forms of schooling, Kristina wants her teaching to be culturally responsive if not liberating. In that sense, Kristina is in good company. In the scholarly fields of critical race theory, critical pedagogy, antiracist teaching, and feminist pedagogies, the observation that public schooling in the United States, without efforts to the contrary, tends to default to mainstream White, middle-class, male, and heteronormative methods and cultural expectations is a foundational assertion. Identified theoretically (Foucault, 1995; Freire, 1993; Gore, 1993; hooks, 1994; Kumashiro, 2000; Ladson-Billings, 1999; Leistyna, Woodrum, & Sherblom, 1996; McCarthy and Crichlow, 1993; Milner, 2003; Solórzano & Yosso, 2002) and observed empirically (Banks, 2002; Cochran-Smith, 2000; Dillabough, 2003; Ferguson, 2003; Fordham, 1988; Grant & Sleeter, 1986; Hall, 2006; Hallinan, 2001; McCarthy & Crichlow, 1993; Noguera, 2003; Ostrove & Cole, 2003; Skiba, Michael, Nardo, & Peterson, 2002; Solórzano & Yosso, 2002; Steele, 1997; Suzuki & Aronson, 2005), this assertion is seldom challenged by any of the most respected scholars of education in this country.

Though Kristina understands math education to be a tool for the enhancement of equity, as a White woman, she is afraid her identity "speaks for itself" in terms of generating a racialized and gendered distrust among her boys of color. In these moments, Kristina's internal understanding of her identity as a politically aware teacher seems to collide with her students' (possible) perception of her as just another "White nagging woman." Although she wants not to be perceived in this manner, if she confronts the Black male students about their off-task behavior, she feels as though she has no choice but to participate in that identity and the process of "White education." This apparent no-win situation makes her feel paralyzed such that she does not "know how far to push" and becomes "insecure."

For Kristina, her insecurity about what to do and her vulnerability about how she feels may emerge from suddenly losing the privilege of being viewed as an individual (something she possesses by virtue of her Whiteness) and instead being seen as the personification of a negative stereotype (something with which the Black males may be very familiar). Facing the stark differences between how students perceive her and how she perceives herself foments a sort of *paralysis of practice*, which is an inability to discern the proper course of action or even how to think about what one is experiencing.

This paralysis of practice is something that non-White teachers seem to experience as well. In the following exchange between a Project ASSERT moderator and Elsa, a late-twenties, Peruvian social-studies teacher who is also a lesbian, note how the teacher struggles with the identity-perception gap:

Elsa: I have one class where I only have two boys. And, only one person is not Black, and she's White. And, a couple of times she has been absent. She's brought in excuses. She's been very good about her work. And I heard some of the girls mutter "[The teachers] let her do that 'cause she's White." I have never thought that to myself, because I am not White so I am not supposed to have that kind of bias. But what's being really difficult for me right now is, as much as I can think of instances like that, I don't know how to pull them together. I don't know how to connect them. And, if I can't connect them, I don't really know what I can do in the group with things like that. These flashes. These instances.

Moderator: When you say "connect them"?

Elsa: I can't find a pattern. I can't find a way to bring them to the table so that we can discuss [these instances] and me feeling that I can do something with it. I think that it is awful that I'm caught in a situation where I hear something like that. Well, maybe that's where I should start—how do I react when I hear something like that? Because it was a very private, very soft, very bitter reaction. Do I put the [White] girl on the spot? Do I want to open it up and talk about it? She's already a huge minority in this room, and she didn't seem to hear it. Do I want to open up a can of worms if it's really happening? And if it's only this, these two people here that are complaining, how is it affecting her? Do I want to ask her? And plain in her face? Maybe that's more where my interests would be. But, in terms of just bringing, these are things that I've heard that sounded racist, or these are things that I've seen the kids do that were racist. I don't know what to say.

To face the possibility that she, as a teacher of color, may be perceived to be racist in the way she allocates homework extensions is a difficult prospect for Elsa, particularly in light of her background in activism and coursework in critical pedagogy. If she is to listen to her students and take their perceptions seriously, she has to consider the possibility that she and/or her practices are perceived as racist. The fact that a student would name Elsa's actions as evidence of racial prejudice calls into question whether Elsa can claim to be devoid of such bias, and this throws her identity as an antioppressive teacher into question.

Bill, a Jewish veteran teacher of English in his early sixties, responds to Elsa's statements by focusing on how he would feel were he in her situation: "I mean, my reaction to it is I just, inside I would be furious. I really would. That they think that about me. So, I know I would confront the two girls privately. That's what I would do, because I can't live with myself unless I do."

When Bill asks Elsa how she felt in that moment, she replies less with outward anger and more with inward castigation. "I feel humiliated," she admits, "that someone says that in my room. That someone thinks something like that of me." She later adds, "It's so humiliating that I don't feel the authority yet. I have talked to the girls, but I don't feel that sense of authority of [being able to say] 'How dare you!' I feel bad. I feel ashamed." Responding to Bill's request for clarification, Elsa explains, "I feel bad, that they would say something like that in my class, that they haven't learned."

Facing the possibility that students believe their actions are examples of favoritism, if not prejudice, provokes strong emotions in Elsa and Bill. Whereas Bill's reaction is anger at being wrongfully (mis)understood, Elsa's is humiliation at being perceived in a manner contrary to her expressed purposes and shame at failing to possess the requisite "authority" to respond to the student's comment. That such emotion could be generated from a student's briefly overheard comment indicates how deeply felt the identity-perception gap can be experienced and how vulnerable teachers often feel when it is revealed. This also suggests that a teacher's awareness of identity-perception gaps may rouse difficult questions about purpose, character, efficacy, and agency.

For Kristina, the identity-perception gap reveals the potential to be labeled, dismissed, and "othered" (i.e., relegated to an outsider status and/or stigmatized as inferior when compared to a presumed norm) by her students, a particularly frustrating and painful possibility. In her teaching, she chooses not to hide her sexual identity from her students and also openly acknowledges her German roots. In doing so, however, she feels as though the students can too easily reject her admonishments about using homophobic language in the classroom because she is perceived to be so markedly different from (some of) them. She states, "It's annoying to me to be reduced to the lesbian math teacher. Once I've been here for a while, I may be the German math teacher . . . the German lesbian math teacher. I'm the tall White lady math teacher, you know?" When describing the actions she takes to address and prevent homophobic language in her classroom and the students' responses to them, it is the process of being "othered" that exposes the identity-perception gap and leaves her vulnerable to dismissal. She explains, "Maybe they do think about [what I've said about homophobic language] and it does make sense and they use [such language] afterwards just to spite me, but maybe it's so easy for me to become the other that it ends up backfiring." Here she is referencing an experience of being seen by her students in a way that makes her so much the outsider that she becomes irrelevant. This marginalization marks her experience as invalid and unworthy of her students' consideration precisely at the moment she is trying to make that experience the reason for changing their behavior. In their dismissal of the validity of her claims about homophobic language, she experiences a rejection of her identity, and this leaves her feeling vulnerable.

When teachers describe confusion about what their "place" is, whether they are "in a position" to say something or not, if it makes a difference "who they are" when they teach across difference, or how to understand what it means to be "a White woman telling [a student of color] what to do," they are referencing their social position relative to the students. This positioning shapes how teachers read students' identities and their own. It is inextricable from relations of power because in a society as hierarchical as ours, position *is* power. Students may read their teachers in accordance with dominant constructions of what a teacher is and does, mainly that they are benign if not helpful guides; or, students may read their teachers more oppositionally, resisting and rejecting the dominant constructions they represent and viewing them instead as an irrelevant nuisance or even as an oppressor. Further still, given the requisite supports, there may be a negotiated reading in which teacher and student engage one another's perspectives as valid and learn to empathize with each other's needs. Regardless of how students read their teachers, once the gap between "who I think I am" and "who students think I am" is revealed to the teacher, the politics of identity and the importance of position become apparent. The extent to which teachers are able to dialogue about such issues may affect their ability to determine effective practices when the identity-perception gap is revealed.

If teachers possess a certain commitment to justice and to antiracist, antisexist, or antihomophobic pedagogy and experience themselves as caring deeply for all students regardless of racial, cultural, gender, sexual, or socioeconomic markers, it can be disarming for these teachers to suddenly confront the fact that such ambitions may have no currency with students who see them only according to their (perceived) social position. To recognize that one's perceived identity impacts the capacity for classroom efficacy is to call into question the teacher's ability to teach. This is why an awareness of the identity-perception gap can have such an impact—it may threaten the teacher's sense of worth. Once the identity-perception gap is exposed, relational dynamics, emotional reactions, and identity politics may displace pedagogy in the mind of the teacher. Without opportunities to unpack the meaning generated in those exchanges, teachers' confusion and vulnerability may influence forms of decision making that foreclose relational connections with students that could lead to deeper understanding of their needs. For this reason, the identity-perception gap reveals not only the fragility of teachers' identities but the necessity to provide support for those facing stark differences between their self-understanding and how they are perceived by their students.

"Safety" and "Comfort" at the Expense of Critique

Closely related to experiences of vulnerability are the ways teachers variously seek comfort when confronting the identity-perception gap. When teachers negotiate meaning and power with students across difference, it can expose a multiplicity of perspectives that challenge the authority of any single view. When that view is the one the teacher possesses of herself or himself, the difference between "who I think I am" and "who students think I am" is revealed, and that gap can be uncomfortable. The manner in which the participants of this study brought such experiences to their peers suggests a desire for comfort that often leads in two different directions: toward a didactic management of differing perspectives, or toward a negation of the critiques implicit in them. That is, the teachers often sought comfort either through the establishment and maintenance of "safe" spaces, or through the rejection of perspectives that were critical of their teaching.

Teachers often face the identity-perception gap when they try to maintain a "safe" classroom in which all are able to learn. In doing so, many teachers overhear such loaded and historically oppressive terms as "nigger," "faggot," and "bitch" as a matter of routine. When students use language that may be comfortable for (some of) them but is experienced as derogatory or inappropriate by the teachers, the identity-perception gap is revealed. How teachers understand and address the gap reveals critical differences between how the teachers see themselves and how the students see the teachers.

In the following exchange, Elsa and Bill struggle to address what they understand to be offensive language, and both express confusion about what to do when students use such terms so casually even when they are experienced as hurtful by others.

> *Elsa*: Whenever I stop the class and try to talk to the kids about what it means when you say "That's okay, I'm straight," or "That's gay," or any of those two scenarios that I can describe when we finish this one, the response I get is always, "Give me a break. We know each other and we're friends and none of this is for real. This is all just play." And I feel offended and I always tell them that and I mean they eventually stop saying it but you really made me question what we're doing. Everyone knows it's inappropriate. But how do we really understand how they interact with each other? How do we, I mean putting myself in that place, I would feel offended no matter what my face shows and no matter what I answer back. You know, I would feel what you're trying to say is being mean. You're being mean and even though I am not going to show it, I hurt. They are so confident in telling you this. "It's not a big deal, we're clean. Gimme a break. Please don't send me out of the class. We're playing! It's a JOKE!" And the whole class is saying this. And eventually you're like,

am I making too much of a big deal? Should I just say "Don't say it in class," but not be so big about you're really hurting your peers. It's very offensive when you say that. You know, to what extent do I want to give it that much, that much, enforcement. Do I think that I'm being the one that's superwimpy, or hypersensitive?

Bill: I don't know though. It goes back to what you were saying when Blacks, I've heard them use the word "nigger" between themselves. I want to stop them in their tracks because, do they really understand what the hell that means? What connotations that has? I mean, if somebody, if I hear somebody use the word "kike" around me, I go ballistic. Because that's, that's, the holocaust is what that is. And to me, "nigger" means slavery and all those horrible things, so, I agree with you. I don't understand how they get minimized, these terms.

When Elsa states that she "puts herself in [the student's] place," she seems to be communicating an attempt to understand the students' perspectives on the use of such words and how youth may view adults who try to get them to change their language. Intending to negotiate a safe way of communicating with one another in the classroom, the students express to her that she does not understand them and that her pleas, directives, and admonishments reveal a lack of awareness not only of how *they* see the world but of how they see *her*. Rather then being perceived as she would like to be (i.e., as she sees herself—a teacher intent on naming such injurious language as oppressive in order to make the classroom safe for all participants), she is cast by the students as a nuisance, as someone who fails to appreciate the students' perspectives, and as one who relies on her positional authority to impose strictures on their behavior. When confronted with this gap, Elsa begins to doubt herself and question her methods, wondering, "Am I making too much of a big deal? . . . [Am I] being the one that's superwimpy, or hypersensitive?" In responding to Elsa's experience, Bill expresses similar confusion, which in this case is directed at the students: "I don't understand how they get minimized, these terms." For both Elsa and Bill, the gap in perception is as personal as it is emotional.

As these examples illustrate, attempting to negotiate a shared construction of "comfort" or "safety" when there is a gap between how the teacher perceives herself and how the students perceive her can be tricky business. Shannon expresses this sentiment by saying, "There's a certain discomfort around how up front some of the students can be at such a young age. I don't want to say it's scary or intimidating, but it's uncomfortable." When feelings like this arise, teachers necessarily confront the fact that somebody's version of "comfort" will be marginalized when attempts are made to address the discomfort of someone else. If the teacher seeks an experience of "safety" that requires the policing and disciplining of student language, it is possible that the students' perception of that intervention may actually undermine their experience of the classroom as "safe." Like the vulnerability issues explored above, this can leave the teacher feeling stuck.

When students use presumably derogatory language to refer to one another, several of the teachers in the study are reticent to accept such words as terms of endearment even as they remain confused about how to intervene. Uncomfortable allowing students to claim such words are functioning as affection, Shannon explains, "I think that the kids, in spite of what they say about 'Oh, the teachers take things too seriously,' they'd like to have this freedom, I think, to kid around with one another, 'cause they say, 'We're just playing. I'm just playing. This is my friend.'" Asked about her perspective on language versus the students' perspective, Shannon then remarks, "I asked the females I was teaching, 'You allow the guys to call you "bitches." Why is that? Doesn't that offend you?' And so they said, 'No, but if they called us a skank, it would.' And, you know, it was taking it at a different level . . . To them, they thought it was a term of endearment . . . To me, I think it's awful." In these excerpts, Shannon first maintains that the school's guidelines restricting student language make the classroom "safe," then she asserts that such guidelines miss the point of what the students consider "comfortable." Again, the gap in perception leads to confusion, insecurity, and an inability to act decisively.

Struggling to find a cohesive set of criteria to govern when to intervene and when not to, teachers like Shannon necessarily confront the identity-perception gap. To negotiate with students which words will be deemed reprehensible and which ones will not exposes how people with different identities have different needs of language. Told not to say something that is culturally acceptable within one's peer group, the student may reject the teacher's interventions and call into question her ability to understand students or her desire to hear youth on their own terms. Likewise, the teacher, upon hearing these words and the rationalizations for their use, may be utterly dumbfounded at how anyone, especially the students he is trying to prepare for adulthood, could ever use them in a supposedly benign fashion. This can result in confusion about exactly what should be negotiable and where limits should be set. Commenting on this challenge, Shannon remarks, "Where do you draw the not-giving-the-attention, giving-the-attention [line]? . . . I mean, we came up with our guidelines and you follow a rubric of the classroom management, but it's gray. And most of it's gray area. And it's difficult." It may be perceived as "gray" and "difficult" partly because the meaning and impact of the words are relationally constructed, open to multiple interpretations, provisional, and contingent to context and participants. Who gets to name the authoritative meaning of such words has everything to do with how power is constructed in the identity-perception gap.

As case in point, the teachers are asked in a separate conversation if, when confronted with misogynistic language, they name it as such with their students. Shannon responds, "I don't think so. I mean, I will label it in terms of the safety language [of the school] . . . but I don't say 'This is demeaning,' or 'This is disrespectful of women.' I mean, I don't even feel like I could say that . . . [The reason I'm not going to do that is because] I'm not going to get into a fight with the students [about] whether this was a gender thing or not. I'd rather keep it in general terms and move on with my learning." Elsa then responds, "I was thinking exactly the same thing. I was thinking, okay, if we do it the other way, then you're opening up a whole other topic that everybody can argue about. And depending on what day it is, do you want to go there or do you want to keep going with your equation or your [Great] Depression timeline?" This suggests that both teachers are reluctant to negotiate meaning or "fight" with students who may have alternate interpretations of and experiences with the language being used. To open that line of inquiry would be to expose a contentious gap in perception, and this teacher seems to indicate it is more important to move ahead with the lesson and maintain a specific "comfort level" than it is to explore what those gaps may reveal. It also suggests that the need to teach the formal curriculum (i.e., the content mandated by state, district, and/or school frameworks) trumps the need to address the social inequities hidden by the language being privileged in the classroom. That the teachers themselves indicate a lack of preparedness in addressing such language suggests that they may have had little to no practice doing it themselves.

Sometimes, when a teacher attempts to comfort another teacher who is experiencing anxiety or confusion as a result of a student's comment, that comfort comes at the expense of the student's perspective. In an apparent attempt to alleviate Elsa's feelings of humiliation and shame regarding the accusation that she favors her one White student over others, Julio delivers several statements that effectively reject the legitimacy of the student's perception. To Julio, the student's remark is evidence of "this typical stereotypical attitude that they have" and is not a valid critique of her teaching. Reacting to how other teachers were framing Elsa's interaction with her students, he says, "I'm going to object to the psychological part of this to you. And I think it's totally rationalization on [the students'] part. They're using you, you know? They're making you the enemy. They're, you know, very typical kind of stuff." Julio later claims that, compared to elementary-school students, high-school students simply are not as concerned with learning about differences between people. Explaining this seemingly indefensible assertion, Julio posits, "At this level, they're not curious. They might be curious at some point, but they kind of feel like they know more than you and I." Believing the students' perspectives carry little validity compared to the teachers' perspectives, Julio explains that his approach would be to "just try to find a way where I can get those kids to understand that that's just not true" if ever confronted with accusations that his teaching practices were prejudiced. The implication here is that because teachers are understood to have the correct knowledge of the situation, the students' knowledge can be devalued, if not altogether rejected. These descriptions suggest that, when faced with the identity-perception gap, teachers may demonstrate little openness to the possibility that students' perceptions may have merit. They seem to possess multiple ways of dismissing students' perceptions and may use them to support other teachers who feel uncomfortable with the critiques implicit in the identity-perception gap.

This cessation of critique seems to extend into the way teachers react to differences in identity-perception among one another too. Speaking to a fellow teacher, Shannon remarks, "For me, anyway, to feel safe, to feel like I'm going to go really deep and I'm going to get really challenged, I need to feel very comfortable with all of you like I really know you and I trust that you're going to understand, you're not going to judge me for what I'm saying." One interpretation of this plea is that she wants the benefit of the doubt and a reasonable modicum of gentleness if she is to be pushed by her peers. Another interpretation is that she desires to be understood on her terms, to be seen only the way she sees herself, and to keep the identity-perception gap unexposed and unexamined. She admits at one point, "I don't want to be challenged by people that don't know me." If this teacher and those like her expect the same conditions in their classroom discussions with students, it is clear that the desire for comfort and safety may come at the expense of deeper discussions about who we are and how we relate across difference.

These data suggest that hard and fast rubrics about how to make students "comfortable" or how to make a classroom "safe" may only underscore for students how little agency they possess in shaping the discourse. The teacher who reprimands or lectures the student who uses contentious or offensive terms may actually be perceived as an oppressor rather than a liberator even when the teacher's intentions may be noble. While this didactic approach of establishing rules and sanctions may effectively protect both the teacher and students from hurtful exchanges, it may also institutionalize an avoidance of the identity-perception gap and a cessation of healthy critique. If classroom "safety" means no one is allowed to name things as discriminatory, sexist, homophobic, or racist or to talk about how those perceptions may shift according to who one is, then that version of "safety" may actually be counterproductive. For example, Bill claims he uses "the word 'polite' [with students] all the time" and admits that they "get sick of me using it . . . But I do it over and over again, and I think it works when you pound it into them." Is it truly "safe" for the students if the way the teacher contains the discourse is mandated rather than negotiated, if one must "pound it into them"? Such examples suggest that even teachers who remain committed to the promotion of equity and diversity may be perceived as too authoritarian if they foreclose opportunities to discuss how language is perceived differently by different people. That teachers need practice in having such conversations and in negotiating meaning in the identity-perception gap is clear.

Conclusion: The Identity-Perception Gap Is a Resource, Not an Impediment

Working with students from diverse backgrounds inevitably forces teachers to relativize their experiences and question their perceptions. This is partly why Parker Palmer insists that teaching is "a daily exercise in vulnerability" (1998, p. 29). In the confrontation with difference, teachers seem to readily access the emotional elements of their experience as well as the cultural, gender, racial, sexual, and socioeconomic aspects of how they understand it. The presence of vulnerability and heightened affect in these interactions suggests something is at stake—something precious. More often than not, that "something" is the teacher's identity, wrapped up as it is in relationships with students and practices ostensibly designed to serve their needs. Indeed, it can be painful when a teacher works so hard to help students or even goes so far as to identify with the struggles they face when those same students malign, reduce, or (mis)understand the teacher's identity or motivations. When anxiety or anger surface at the moment the identity-perception gap is confronted, this is a clue that something important is going on. The affect is there because the risks are high. To entertain the possibility that the students' contrary or critical view of the teacher may possess some validity requires an immersion in vulnerability and the potential for humiliation and shame—a difficult set of experiences to confront without support and dialogue. Consequently, the competing approach—to denigrate the students' perceptions as being invalid or reprehensible or to deny their salience altogether—often wins out. Understood as a threat in such cases, the many resources intrinsic to the identity-perception gap are too easily squandered.

But what if the identity-perception gap were engaged not as an impediment to understanding but as an opportunity for growth? What if teachers were provided with opportunities to discuss with colleagues the vulnerabilities exposed by the identity-perception gap and to explore what it may reveal about teachers and teaching? Further still, what if teachers were shown how they might discuss the identity-perception gap with the students themselves? As these teachers' conversations and comments show, the identity-perception gap is rich with insight, worthy of regular investigation, and ultimately unavoidable. Ellsworth concurs that "the eruptive, unruly space between" a teacher's identity perception and the student's will not "go away because it's populated by the difference between conscious and unconscious knowledges, conscious and unconscious desires" (1997, p. 41). As long as classroom relations are shaped by broader social antagonisms and interpreted through our unconscious, teachers cannot foreclose the space of difference between how they see themselves and how they may be seen by their students (p. 41).

This begs the question: Why would teachers ever want to confront the impossibility, unpredictability, and anxiety of the identity-perception gap when what they face each day is already often too much? If a teacher's students really just need to pass the upcoming high-stakes test, why spend time and energy on things that may only reveal pain and discomfort that may make the teacher feel less secure, more vulnerable, and unsafe? It is true that if we allow teachers to persist in the myth that education is really about either "getting it" or "not getting it," then dwelling on the identity-perception gap and the anxieties it produces makes little sense. However, if we realize that teacher-student relationships often get stuck and learning stagnates precisely when the student (or the teacher) does not *want* to "get it" or, when she or he does "get it," does not *desire* it (Ellsworth, 1997, p. 46), then we can see that interrogating the identity-perception gap is all about effective practice, academic performance, equity, and diversity.

Therefore, this study suggests that a teacher's unresolved self that gets exposed by the identity-perception gap should not be approached as a problem to solve or a loss to grieve; instead, it might be better framed as an opportunity for a deeper and more complex understanding of one another. This may mean pushing against the predominant manner in which teachers frame their vocational identities. Janet Alsup (2006), for example, contends that teachers are often prepared to believe "that they should not reveal their personal ideologies or make pedagogical decisions based on their racial, ethnic, or gender subjectivities; on the contrary, and in order to be fair to all students, they should be intellectually neutral (and, of course academically rigorous) as often as possible. However, this suppression of personal identity is only a sham, a faćade, because personal subjectivities and ideologies do not disappear; they simply remain, and even fester, as sites of tension and discomfort" (p. 41).

Without risking one's perceptions in relationships and practices, the identities teachers construct may become inflexible or incapable of withstanding the complexities required to relate to one another in diverse settings. When teachers are brave enough to face the identity-perception gap, they need to be discouraged from trying to overcome it because doing so establishes as the goal the attainment of a correct perspective (often achieved by a lack of attention to other's perceptions and needs), which silences dissent and offers a severely reduced, if not entirely false, understanding of teacher-student relationships across difference. Properly conceived, the identity-perception gap offers opportunities for deep engagement with *who teachers are* as the educators of their specific youth. Viewing the gap as a resource puts teachers and students in relationship with each other's intersubjective experiences so that they may validate the authenticity of each, giving both parties the agency they need to learn. This may eventually allow for a deeper appreciation of diversity to flourish.

Both students and teachers alike make decisions about how hard they will work, the manner in which they will behave, what they will believe, and who they will trust based on their perceptions of other people's identities. It makes sense, then, that teachers and teacher educators should look for moments in which the stability of teacher identities is shaken in encounters with students. Calibrating our pedagogical radars to be sensitive to vulnerability and suspicious of efforts to provide comfort or safety will help teachers and teacher educators to detect the identity-perception gap and support others doing the same. In such moments educators may ask of one another: How might the students' perceptions of us be valid? What about the students' perceptions are we reacting against? What would we have to change in our teaching if the students' perceptions were at least partly valid? Once we have explored this issue together, what needs to happen for us to feel stable again and to confidently promote all students' academic achievement?

To be able to see what the identity-perception gap reveals—not that one perception is right and the other is wrong but that both may function as real in the minds of each person—is to enter the classroom with a profound appreciation for what is at stake in teacher-student relationships. Teaching and learning are dependent on a confrontation with something new, something different, and an active engagement with difference is a primary mechanism for developing our capacity to think and act with complexity. It follows then that gaps in perception may alert us to the possibility for greater understanding among people differently positioned in society. After all, it is hard to appreciate diversity when the perceptions of others are experienced as threats, and it is unlikely that a teacher will strive for equity for all students when some are cast as a menace. In the end, the achievement of equity and the enhancement of diversity depend on our capacity not just to hear others but to relate to them as if their perceptions are valid.

References

Alsup, J. (2006). *Teacher identity discourses: Negotiating personal and professional spaces.* Mahwah, NJ: L. Erlbaum Associates.

Banks, J. A. (2002). Race, knowledge construction, and education in the USA: Lessons from history. *Race Ethnicity and Education, 5*(1), 7–27.

Bernstein, B. (2000). *Pedagogy, symbolic control and identity: Theory, research, critique.* New York: Rowman & Littlefield.

Bingham, C. W. (2001). *Schools of recognition: Identity politics and classroom practices.* Lanham, MD: Rowman & Littlefield.

Cochran-Smith, M. (2000). Blind vision: Unlearning racism in teacher education. *Harvard Educational Review, 70*(2), 157–190.

Dance, L. J. (2002). *Tough fronts: The impact of street culture on schooling.* New York: RoutledgeFalmer.

Dillabough, J. (2003). Gender, education, and society: The limits and possibilities of feminist reproduction theory. *Sociology of Education, 76,* 376–379.

Ellsworth, E. (1997). *Teaching positions: Difference, pedagogy, and the power of address.* New York: Teachers College Press.

Erikson, E. H. (1968). *Identity, youth, and crisis.* New York: W. W. Norton.

Ferguson, A. A. (2000). *Bad boys: Public schools in the making of Black masculinity.* Ann Arbor, MI: University of Michigan Press.

Ferguson, R. F. (2003). Teachers' perceptions and expectations and the Black-White test score gap. *Urban Education, 38*(4), 460–507.

Fordham, S. (1988). Racelessness as a factor in Black students' school success: Pragmatic strategy or pyrrhic victory. *Harvard Educational Review, 58*(1), 54–84.

Foucault, M. (1995). *Discipline and punish: The birth of the prison* (A. Sheridan, Trans. 2nd ed.). New York: Vintage Books.

Freire, P. (1993). *Pedagogy of the oppressed.* New York: Continuum.

Gore, J. (1993). *The struggle for pedagogies: Critical and feminist discourses as regimes of truth.* New York: Routledge.

Grant, C. A., & Sleeter, C. E. (1986). Race, class, and gender in educational research: An argument for integrative analysis. *Review of Educational Research, 56*(2), 195–211.

Hall, H. R. (2006). Teach to reach: Addressing lesbian, gay, bisexual, and transgender youth issues in the classroom. *The New Educator, 2,* 149–157.

Hallinan, M. T. (2001). Sociological perspectives on Black-White inequalities in American schooling. *Sociology of Education, 74*(Extra Issue), 50–70.

hooks, b. (1994). *Teaching to transgress: Education as the practice of freedom.* New York: Routledge.

Kumashiro, K. K. (2000). Toward a theory of anti-oppressive education. *Review of Educational Research, 70*(1), 25–53.

Ladson-Billings, G. (1999). Just what is critical race theory and what's it doing in a nice field like education? In L. Parker, D. Deyhle, & S. Villenas (Eds.), *Race is . . . race isn't.* (pp. 7–30). Boulder, CO: Westview Press.

Leistyna, P., Woodrum, A., & Sherblom, S. A. (1996). *Breaking free: The transformative power of critical pedagogy*. Cambridge, MA: Harvard Educational Review.

McCarthy, C., & Crichlow, W. (1993). *Race, identity, and representation in education*. New York: Routledge.

Milner, H. R. (2003). Reflection, racial competence, and critical pedagogy: How do we prepare pre-service teachers to pose tough questions? *Race Ethnicity and Education*, *6*(2), 193–208.

Noguera, P. A. (2003). The trouble with Black boys: The role and influence of environmental and cultural factors on the academic performance of African American males. *Urban Education*, *38*(4), 431–459.

Ostrove, J. M., & Cole, E. R. (2003). Privileging class: Toward a critical psychology of social class in the context of education. *Journal of Social Issues*, *59*(4), 677–692.

Palmer, P. J. (1998). *The courage to teach: Exploring the inner landscape of a teacher's life*. San Francisco: Jossey-Bass.

Skiba, R. J., Michael, R. S., Nardo, A. C., & Peterson, R. L. (2002). The color of discipline: Sources of racial and gender disproportionality in school punishment. *The Urban Review*, *34*(4), 317–342.

Solórzano, D. G., & Yosso, T. J. (2002). Critical race methodology: Counter-storytelling as an analytical framework for education research. *Qualitative Inquiry*, *8*(1), 23–44.

Steele, C. M. (1997). A threat in the air: How stereotypes shape intellectual identity and performance. *American Psychologist*, *52*(6), 613–629.

Suzuki, L. & Aronson, J. (2005). The cultural malleability of intelligence and its impact on the racial/ethnic hierarchy. *Psychology, Public Policy, and Law*, *11*(2), 320–327.

Taubman, P. (2002). Facing the terror within: Exploring the personal in multicultural education. In C. Korn & A. Bursztyn (Eds.), *Rethinking Multicultural Education: Case Studies in Cultural Transition* (pp. 97–129). Westport, CT: Bergin & Garvey.

2

Developing a Multicultural Curriculum in a Predominantly White Teaching Context

Lessons from an African American Teacher in a Suburban English Classroom

H. Richard Milner IV, Vanderbilt University

> You teach what you know; you teach what you've experienced; you teach who you are. And when we have White teachers who don't deal with race and culture and difference, it's really a handicap to the students because they are not teaching reality. My students know me. They know how I live, and there's no misunderstanding, no misinterpretations about that. I am a Black woman, and they need to understand that there are some differences between myself and them . . . My experiences aren't exactly like theirs and part of that has to do with the fact that I am Black. Racism does exist; it existed decades ago, and we're still grappling with it.
>
> —Dr. Wilson, the teacher in this study

In this study, I sought to understand an African American teacher's curriculum development and teaching in a suburban (mostly White) high school. (The terms "African American" and "Black" are used interchangeably throughout this chapter.) The study contributes to the literature in that it focuses on a context that might otherwise be ignored because there is not

---

This chapter has been reprinted/adapted from Milner, H. R. (2005). Developing a multicultural curriculum in a predominantly White teaching context: Lessons from an African American teacher in a suburban English classroom. *Curriculum Inquiry, 35*(4), 391–427.

a large student-of-color representation in the school. Evidence is needed to understand how multicultural education emerges and is received in a variety of contexts, including suburban schools, because students will interact with multicultural, multiracial, multilingual, and multiethnic people in the United States for the remainder of their lives. In like form, these people of color will interact with mainstream students, and it is essential for all students to develop knowledge, skills, awareness, and understandings in order to live healthy and productive lives. In addition, research agendas are critical concerning the effectiveness of multicultural curricula in White contexts because "both children of color and White children develop a 'White bias' by the time they enter kindergarten" (Banks, 1995, p. 392). These biases and assumptions need to be problematized in order to help students develop more appropriate lenses for thinking about knowledge, self, others, and society.

Teaching for Racial and Cultural Awareness

Researchers and theoreticians agree that multicultural education is both a necessity and an advantage for academic and social success among students of color (Banks, 1998; Ford, 1996; Grant & Tate, 1995). Students of color need to encounter and experience a curriculum that highlights, showcases, and speaks from the point of view of the life experiences and contributions of people of color, women, and other marginalized groups—not just those of the White mainstream. In considering the importance of students' cultural and social experiences in the curriculum, Ladson-Billings (1994) explained that teachers can maximize student learning by "importing the culture and everyday experiences of the students" (p. 117). Banks (1998) maintained that a significant goal of multicultural education was to "increase educational equality for both gender groups, for students from diverse ethnic and cultural groups, and for exceptional students" (p. 22). Jenks, Lee, and Kanpol (2001) suggested that additional goals of multicultural education involved "knowledge of cultural and racial differences and issues; the critical examination of one's own beliefs and values regarding culture, race, and social class; and an understanding of how knowledge, beliefs, and values determine one's behavior" (p. 88).

Ensuring that various cultural, racial, ethnic, gendered, and linguistic groups of people and their experiences are represented in the curriculum is not the only essential feature in providing access, empowerment, and awareness for students of color. The very *nature* of this *content* and *how* it is actually incorporated into the lessons are also critical. Gay (2000) asserted that students often felt "insulted, embarrassed, ashamed, and angered when reading and hearing negative portrayals of their ethnic groups or not hearing anything at all" (p. 116). Thus, it is not enough to incorporate the historical, political, and social experiences, events, and challenges of various ethnic groups into the curriculum. But rather, the nature of that curriculum content (what is actually included, how, and why) is very important as students come to understand themselves and others in a pluralistic society.

Multicultural Curriculum for Whom?

One of the recurring questions that emerges in the literature is for whom multicultural education is intended and needed. That is, in what types of schools, with what population of students, and by what means ought multicultural curricula and instruction emerge? Understandably, because the number of culturally, racially, and linguistically diverse students in the United States is increasing at a high rate, and because the number of teachers remains largely White, middle class, and female, studies have consistently explored the effectiveness of White teachers teaching in predominantly African American or other ethnic settings (see Cooper, 2003; Johnson, 2002; Ladson-Billings, 1994; Rushton, 2004). This important line of research helps us to think about and to conceptualize some of the complexities inherent in the cultural and racial disconnections, connections, and mismatches that often emerge in classrooms between teachers and students. Clearly, as Foster (1995) and Smith (2000) explained, teachers' backgrounds (cultural, racial, linguistic, and ethnic) play critical roles in the education of students, particularly students of color. The idea is that "some groups of students—because their cultural characteristics are more consistent with the culture, norms, and expectations of the school [and their teachers] than are those of other groups of students—have greater opportunities for academic success than do students whose cultures are less consistent with the school [and teachers'] culture" (Banks, 1998, pp. 22–23). Critics, however, suggest that multicultural education can divide society and the world. Ladson-Billings (1992a) explained in response to this pervasive critique that multicultural education "advocate[s] unity *and* diversity" (p. 310). In other words, "cultural differences do not [and should not] necessarily result in lack of unity" (Ladson-Billings, 1992a, p. 308). Rather, multicultural education can help build bridges and to unite diverse groups from various life worlds.

Researchers recognize that teachers are themselves curriculum developers—they are more than mere curriculum implementers (Delpit, 1995; Gay, 1994, 2000; Ladson-Billings, 1994; McCutcheon, 2002; McCutcheon & Milner, 2002; Milner, 2003a; Milner & McCutcheon, 2002). This study is different in that it investigates how a Black teacher attempts to develop the curriculum and incorporate multicultural curriculum *into a predominantly*

White learning context. The study embraces Banks's (1998) notion that multicultural education "is to help *all* students, *including White mainstream students*, to develop the knowledge, skills, and attitudes they will need to survive and function effectively in a future U.S. society in which one out of every three people will be a person of color" (p. 23, emphasis added). Multicultural education, then, is not only concerned with students of color and linguistically diverse learners. It is also focused on students in the mainstream of learning. However, as Jenks et al. (2001) put it, "In suburban schools in which the population is basically white and middle-class, multicultural education is often viewed as unnecessary" (p. 87), and teachers, community members, and parents often adopt color-blind ideologies and philosophies (Johnson, 2002; Lewis, 2001; Milner, in press; Milner, 2003b; Milner & Smithey, 2003) in the development of curriculum and in instruction.

The research is clear that when teachers do not "see color," or at least acknowledge that race matters, there may be "ignored discriminatory institutional practices toward students of color such as higher suspension rates for African American males" (Johnson, 2002, p. 154) in conjunction with students of color being referred to special education and lower tracked courses in general. Banks (2001) explained that "a statement such as 'I don't see color' reveals a privileged position that refuses to legitimize racial identifications that are very important to people of color and that are often used to justify inaction and perpetuation of the status quo" (p. 12). Moreover, as Ford (1996) explained, "Black students, particularly males, are three times as likely as White males to be in a class for the educable mentally retarded, but only half as likely to be placed in a class for the gifted. Not only are Black students under enrolled in gifted education programs . . . [but] Black students are over-represented in special education, in the lowest ability groups and tracks, and among high school and college dropouts" (p. 5).

The idea that racial discrimination and cultural misunderstandings do not exist in predominantly White settings is a fallacy. Further, students who attend these mostly White settings do not live in a vacuum; they will experience diversity in the world, and they must be prepared in order to function effectively in the world. In Banks's (1995) words, "Multicultural education . . . if implemented in thoughtful, creative, and effective ways, has the potential to transform schools and other educational institutions in ways that will enable them to prepare students to live and function effectively in the coming century" (p. 391).

Integrating Multicultural Content

Research and theory are clear that teachers in public schools and at the college level must rethink, renegotiate, and transform the nature of their curriculum from more traditional models, where many ethnic groups are either misrepresented or not represented at all, to curricula that are more inclusive of different racial, linguistic, and cultural groups (see, for instance, Arias & Poynor, 2001; Banks, Cookson, & Gay, 2001; Dillard, 1996; Troutman, Pankratius, & Gallvan, 1999). Models and typologies exist, serving as heuristics, in order to help researchers and theoreticians think about the stages of development among teachers in the process of creating multicultural content and instruction to enhance student learning. In short, we have come to understand that developing the skills to effectively create and to deliver multicultural content is a process that must be studied in order to draw conclusions about its procedural effectiveness.

Research and theory have also focused on the development of pedagogical approaches (the how) across various contexts to better meet the needs of students of color and ultimately all students (Gay, 2000; hooks, 1994; Ladson-Billings, 1994; Ladson-Billings, 2000). For instance, as Ladson-Billings (1992b) maintained, culturally relevant pedagogy is an approach that "serves to empower students to the point where they will be able to examine critically educational content and process and ask what its role is in creating a truly democratic and multicultural society. It uses the students' culture to help them create meaning and understand the world. Thus, not only academic success, but also social and cultural success is emphasized" (p. 110).

Ladson-Billings (1994) further explained that culturally relevant pedagogy "uses student culture in order to maintain it and to transcend the negative effects of the dominant culture. The negative effects are brought about, for example, by not seeing one's history, culture, or background represented in textbook or curriculum . . . culturally relevant teaching is a pedagogy that empowers students intellectually, socially, emotionally, and politically by using cultural referents to impart knowledge, skills, and attitudes" (pp. 17–18).

Moreover, culturally relevant pedagogy is an approach that helps "students to see the contradictions and inequities that existed in their local community and the larger world" (Ladson-Billings, 1992b, p. 382). Gay (2000) defined culturally responsive teaching as "using the cultural knowledge, prior experiences, frames of reference, and performance styles of ethnically diverse students to make learning encounters more relevant to and effective for them. It teaches to and through the strengths of these students. It is culturally validating and affirming" (p. 29).

Perhaps the most widely recognized models to help gauge teachers' (both preservice and inservice) stages of development in the process of multicultural curricula were developed by Sleeter and Grant (1994) and Banks (1998), although other models exist (see Grillo, 1998; McAllister & Irvine, 2000). The development of the models take into account that "teachers do not start with the same type of cross-cultural understanding" (McAllister & Irvine, 2000, p. 19), and the models are often used to support teachers' thinking and beliefs as they develop multicultural competencies and understanding. In addition, the models are not always described linearly; it is understood that teachers may not follow some predetermined process of development. Still, each model outlines stages of development that helps us think about how teachers are coming to conceive and represent curricula. Common across these models is that they end with an action-oriented stage in which students and teachers are encouraged to "do something" about inequitable circumstances that they encounter (Grillo, 1998). Sleeter and Grant (1994) referred to this final stage as the "social reconstructionist" level and Banks (1998) referred to this action-oriented stage as the "social action" approach or level. However, helping teachers in preservice programs and in schools get to this ultimate stage of action seems to be a difficult feat, a point that will be addressed again based on the findings of this study.

For the purposes of this study, I employ Banks's (1998) model of curricula integration, as Banks outlined several approaches that can help us understand the complex processes teachers engage as they create and develop multicultural curricula: the contributions approach; the additive approach; the transformative approach; and the social-action approach. I use this model and these stages as a conceptual tool to analyze this teacher's curriculum and her practice in a suburban high school (see Figure 2.1).

According to Banks (1998), the contributions approach is one that is framed by a celebration of ethnic and cultural groups through a consideration of what they have "given" to society. The contributions approach allows teachers and schools to superficially incorporate the involvement of certain groups of people or individuals by exposing students to highlights or hallmarks achieved by those groups. A common example of this type of approach is Black History Month celebrations that highlight the accomplishments of famous Black Americans such as Dr. Martin Luther King, Jr., and Thurgood Marshall, mostly individuals whom teachers and administrators sift through and believe to be "safe" and comfortable contributors. This means that the schools and teachers are able to select individuals and certain accomplishments that often diminish or ignore other, perhaps more radical contributors, such as Nat Turner or Malcolm X. The approach

LEVEL 4

The Social Action Approach

Students make decisions on important social issues and *take action* to help solve some of the social ills and injustices in their school, community, and society.

LEVEL 3

The Transformation Approach

The curriculum is transformed. The structure of the curriculum is changed to enable students to view concepts, issues, events, and themes from the perspectives of diverse racial and cultural groups. "Tough" topics and themes are not avoided. These issues are central to the entire curriculum, not just one week or unit.

LEVEL 2

The Additive Approach

Content, concepts, themes, and perspectives are added to the curriculum without changing its structure (e.g., Black History Month or Native American Awareness Week). That which is safe, politically correct, and less controversial is more likely to be taught and discussed.

LEVEL 1

The Contributions Approach

Focuses on heroes, holidays, and isolated events of cultural and ethnically diverse groups and individuals. Focuses more on what diverse groups have done than who they are and fails to transform and integrate the curriculum to levels of meaning and depth.

Figure 2.1. Banks's approaches to multicultural curriculum reform

Note. Reprinted with permission of the author from *Cultural Diversity and Education: Foundations, Curriculum, and Teaching* (5th ed.), by James A. Banks, 2006, Boston: Allyn and Bacon, p. 62.

is not central or at the core of the curriculum but is on the margins of what students have the opportunity to learn.

Similar to the contributions approach, a second approach, the additive approach, is one in which "cultural content, concepts, and themes are added to the curriculum without changing its basic structure, purposes and characteristics" (Banks, 1998, p. 30). There is little depth inherent in this approach. Teachers may opt to add a lesson, a book, or a film, for instance, to the already established curriculum without delving into thematic threads that bind a more sustained, coherent, and comprehensive curriculum. Banks (1998) reminded us that both the contributions and the additive approaches "are used to integrate cultural content into the curriculum, people, events, and interpretations related to ethnic groups and women often reflect the norms and values of the dominant culture rather than those of cultural communities" (p. 30). In their analyses, Jenks et al. (2001) expanded that "the danger [of the additive approach] is that if the material becomes an official part of the curriculum, it may be given short shrift—or not be taught at all—by teachers who fail to accept its importance" (p. 97). Teachers may claim to not have enough time to cover the material when it is not imbedded in the curriculum.

The transformative approach actually changes the core and the nature of the curriculum by infusing (not just by addition) multiple views and perspectives into the curriculum—so that the curriculum is not representative of only one dominant view or way of experiencing the world. Students are given opportunities to engage in critical thinking and to develop more reflective perspectives about what they are learning. Jenks et al. (2001) posited that "students learn to be reflective, to adopt different perspectives, and to understand how what they are taught—the knowledge that schooling offers—has been shaped historically, ethnically, culturally, and linguistically" (p. 97). Moreover, Banks (1998) declared that "important aims of the transformative approach are to teach students to think critically and to develop the skills to formulate, document, and justify their conclusions and generalizations" (p. 32). The approach pushes students to look with the *head* and the *heart* (Banks, 2003) as they are critically examining issues both inside and outside of the classroom (Freire, 1998; Wink, 2000). Additionally, the transformative approach to multicultural education and curricula "provide a rich context for equity pedagogy because both transformative curricula and equity pedagogy promote knowledge construction and curriculum reform. Transformative curricula and equity pedagogy also assume that the cultures of students are valid" (Banks & Banks, 1995, p. 155)

The social-action approach, a fourth approach, is an extension of the transformative approach. This approach actually is a form of curriculum that allows teachers to facilitate action-oriented projects and activities related to what students learn from multiple perspectives. Banks's (1998) description of this approach is centered on the importance of helping students come "to know, to care, and to act" (p. 32). Further, Banks (2003) maintained that students must "question the assumptions of institutionalized knowledge and . . . use knowledge to *take action* that will make the world a just place in which to live and work . . . When we teach students how to critique the injustice in the world, we should help them to formulate possibilities *for action* to change the world to make it more democratic and just" (p. 18, emphasis added).

Helping all students reach a higher level of consciousness and awareness can enable their success as they interact with their peers in school as well as others outside of school. Students' behaviors, in this respect, ensure their social transformation as well as academic success.

This teacher is operating at one of the highest levels of Banks's (1998) model (the transformational approach). The study suggests that even teachers highly conscious of race, culture, gender, and ethnicity may find it difficult to reach the highest level of Banks's model, the social-action approach. Clearly, it is difficult to guide students' actions once they have left the classroom. Teachers can encourage students to engage in action-related activities while in their classrooms but cannot control students' behaviors once they have left the classroom. Thus, the transformational stage may be as far as teachers can realistically change the curriculum—that is, while the students that teachers teach are in schools. Ultimately, it is left up to students once they have acquired new and expanded knowledge about social (in)justice issues to change or modify their behaviors once they have graduated and are in the real world. Whereas the transformational stage is cognitive in nature, the social-action stage is more behavioral and sociopolitical. Perhaps helping students reach a level of cognitive transformation through exposure to such a curriculum will help ensure that they will go on to act and facilitate change in society, to speak out against injustice, and to accept, as well as embrace, different ethnic, cultural, linguistic, and racial groups. The idea is that once students *know* better, they are more likely to *do* better.

But what is it about Dr. Wilson and her racial and cultural background that enables her commitment, competence, and success in developing and teaching a cultural curriculum in a predominantly White context? That is, what does a Black teacher bring into a teaching situation that helps her or him develop a transformative curriculum and instruction that her students embrace and that is obviously meaningful?

Black Teachers and Developing the Curriculum

There is a growing and meaningful body of literature that focuses on Black teachers and their teaching. This literature is conceptualized in several important ways: it spans the pre-desegregation era to the present and focuses on P through 12 schools as well as higher education. I shift the discussion at this point to consider this meaningful body of literature. Understanding Black teachers and how they conceive of and represent the curriculum is especially important in this research because Dr. Wilson is a Black teacher whose identity, experiences, and history meaningfully shape her curriculum development and teaching. Dr. Wilson embodied many of the characteristics of the Black teachers explored in this literature, with one noticeable difference—much of the research and theory about Black teachers and teaching focus on Black teachers and their impact in predominantly Black settings. Dr. Wilson teaches in a predominantly White teaching context.

To illuminate, much has been written about Black teachers and their teaching in public-school classrooms (Foster, 1990, 1997; Holmes, 1990; Hudson & Holmes, 1994; Irvine & Irvine, 1983; King, 1993; Milner & Howard, 2004; Milner, 2003a; Monroe & Obidah, 2004), and this literature is not limited to public schools but also highlights Black teachers' experiences in higher education, namely in teacher-education programs (Baszile, 2003; Ladson-Billings, 1996; McGowan, 2000; Milner, & Smithey, 2003). Black teachers and their mission and vision in education have been captured in some insightful ways. Of central importance in this literature is the question, What does a teacher's racial and cultural background have to do with how he or she develops the curriculum and implements it? As Agee (2004) explained, a Black teacher "brings a desire to construct a unique identity as a teacher . . . she [or he] negotiates and renegotiates that identity" (p. 749) to meet their objectives and to meet the needs of their students.

According to hooks (1994), Black female teachers carry with them gendered experiences and perspectives that have been (historically) silenced and marginalized in the discourses about teaching and learning. Although teaching has often been viewed as "women's work," Black women teachers and their worldviews have often been left out of the discussions—even when race was centralized (hooks, 1994). Similarly, in colleges of education, and particularly preservice and inservice programs, the programs are largely tailored to meet the needs of White female teachers (Gay, 2000), and Black teachers along with other teachers of color (male and female) are left out of the proliferation. Where curricular materials were concerned in her study, Agee (2004) explained that "the teacher education texts used in the course made recommendations for using diverse texts or teaching diverse students based on the assumption that preservice teachers are White" (p. 749). Still, Black teachers often have distinctive goals, missions, and decision-making and pedagogical styles of teaching.

In her analyses of valuable African American teachers during segregation, Siddle-Walker (2000) explained, "Consistently remembered for their high expectations for student success, for their dedication, and for their demanding teaching style, these [Black] teachers appear to have worked with the assumption that their job was to be certain that children learned the material presented" (pp. 265–266).

Clearly, these teachers worked overtime to help their African American students learn; although these teachers were teaching their students during segregation, they were also preparing their students for a world of integration (Milner & Howard, 2004; Siddle-Walker, 1996). Moreover, as Tillman (2004) suggested, "These teachers saw potential in their Black students, considered them to be intelligent, and were committed to their success" (p. 282). There was something authentic about these Black teachers. Indeed, they saw their jobs and roles to exceed far beyond the hallways of the school or their classroom. They had a mission to teach their students because they realized the risks and consequences in store for their students if they did not teach them and if the students did not learn. An undereducated and underprepared Black student, during a time when society did not want nor expect these students to succeed, could likely lead to destruction (drug abuse, prison, or even death).

As Pang and Gibson (2001) maintained, "Black educators are far more than physical role models, and they bring diverse family histories, value orientations, and experiences to students in the classroom, attributes often not found in textbooks or viewpoints often omitted" (pp. 260–261). Thus, Black teachers, similar to all teachers, are texts themselves, but these teachers' pages are inundated with life experiences and histories of racism, sexism, and oppression, along with those of strengths, perseverance, and success. Consequently, these teachers' texts are rich and empowering—they have the potential to help students understand the world (Freire, 1998; Wink, 2000) and some of the complexities of race and racism, for example, in powerfully meaningful ways.

As made evident from this body of literature, these African American teachers still often felt irrelevant and voiceless in urban and suburban contexts—even when the topic of conversation was multicultural education (see, Buendia, Gitlin, & Doumbia, 2003; Ladson-Billings, 1996; Milner & Woolfolk Hoy, 2003; Pang & Gibson, 2001). These experiences are unfortunate given the attrition rate of Black teachers in the teaching force. Black teachers are leaving the teaching profession and quickly (Hudson & Holmes, 1994). And historically, particularly pre-desegregation, the teaching profession was always viewed as an honorable and popular profession for Blacks (Foster, 1997; Siddle-Walker, 1996).

In their classrooms, Black teachers were able to develop and implement optimal learning opportunities for students—yet in the larger school context, they were often ridiculed for being too radical or for not being "team players." As evident in my own research (Milner, 2003a) and this study, a Black teacher can feel isolated and ostracized because that teacher often offered a counterstory or counternarrative (Ladson-Billings, 2004; Ladson-Billings & Tate, 1995; Morris, 2004; Parker, 1998; Solorzano, & Yosso, 2001; Tate, 1997) to the pervasive views of their mostly White colleagues. Black teachers' ways of connecting with their students were effective—yet inconsistent with their non-Black colleagues.

For instance, Delpit (1995) shared a reaction from a White teacher when talking about the management style and pedagogical approach of a Black teacher: "It's really a shame but she (that Black teacher upstairs) seems to be so authoritarian, so focused on skills and so teacher directed. Those poor kids never seem to be allowed to really express their creativity. (And she even yells at them)" (p. 33).

What the teacher in the passage above failed to understand was that the "Black teacher upstairs" may have been quite productive and effective in providing all students access and opportunities to learn. Black teachers may have a different way of thinking about how best to make learning happen in their classrooms; in essence, pedagogical and curricula decisions are racially and culturally mediated. They depend on the context, and they are not neutrally constructed.

Irvine (1998) described the interaction between a Black student and a Black teacher borrowing James Vasquez's notion of "warm demanders"; the following quote is from Irene Washington, an African American woman and teacher of twenty-three years: "'That's enough of your nonsense, Darius. Your story does not make sense. I told you time and time again that you must stick to the theme I gave you. Now sit down.' Darius, a first grader trying desperately to tell his story, proceeds slowly to his seat with his head hanging low" (Irvine, 1998, p. 56).

"Warm demander" is a description given to teachers of color "who provide a tough-minded, no-nonsense, structured, and disciplined classroom environment for kids whom society has psychologically and physically abandoned" (p. 56). An outsider observing the teacher's tone and expectations for Darius might frown upon the teacher's approach. However, this teacher's classroom management and pedagogical approach are grounded in a history and a reality that is steeped in care for the student's best interest. In short, the teacher understood quite deeply that she must help Darius learn, and she must "talk the talk." There is a sense of urgency not only for

Irene to "teach her children well but to save and protect them from the perils of urban street life" (p. 56).

Indeed, teachers play an integral role in providing safe, reassuring, and optimal learning for their students. Care is critical for effective learning to occur in any environment. Irvine (2003) explained that teaching is about establishing and maintaining caring relationships, and it is about what Collins (1991) called "other mothering" (and I would add "other fathering"). Black teachers seemed to embrace the idea that their students were like their own biological children, and they wanted the very best for all their students.

Mitchell (1998), in her qualitative study of eight recently retired African American teachers, reminded us of the insight among teachers that can help us understand the important connections between the affective domain and student behavior. Mitchell explained that in order for teachers to establish and to maintain student motivation and engagement, they must be aware of the affective domain. Students' feelings and emotions matter in how they experience education. The teachers in her study "were critically aware of the experiences of the students, both in and out of school, and of the contexts shaping these experiences" (p. 105). The Black teachers in the study were able to connect with the students in the classroom because they understood the students' out-of-school experiences. Thus, these retired Black teachers understood the connection between the home and school, and they were able to conceptualize how students' feelings had been impacted by their home circumstances and consequently how students' feelings emerged in their respective learning in the urban classroom.

More than anything, Siddle-Walker (2000) concluded that, because of the hard work and dedication of Black teachers, "students did not want to let them down" (p. 265). The students put forth effort and achieved academically and socially because "teachers held extracurricular tutoring sessions, visited homes and churches in the community where they taught, even when they did not live in the community, and provided guidance about 'life' responsibilities. They talked with students before and after class, carried a student home if it meant that the child would be able to participate in some extracurricular activity he or she would not otherwise participate in, purchased school supplies for their classroom, and helped to supply clothing for students whose parents had fewer financial resources and scholarship money for those who needed help to go to college" (p. 265).

In sum, Irvine (1998) outlined several important practices among African American teachers and their curricula and pedagogical styles:

> They perceive themselves as parental surrogates and advocates for their African American students. They employ a teaching style filled with rhythmic language and rapid intonation with many instances of repetition, call and

> response, high emotional involvement, creative analogies, figurative language, gestures and body movements, symbolism, aphorisms, and lively and often spontaneous discussions. They use students' everyday cultural and historical experiences in an effort to link new concepts to prior knowledge. They spend classroom and non-classroom time developing a personal relationship with their children, and often tease and joke with their students using dialect or slang to establish this personal relationship. They teach with authority. (p. 57)

At any rate, teachers' in-school and out-of-school experiences—their autobiographies and social realities—influence their curriculum development and teaching with students. In other words, we know that teachers do more than go into a classroom and robotically teach a set of information or materials. Rather, what happens to teachers in their daily lives and experiences (in the supermarket or in a car dealership, for instance) often show up in the curriculum and their teaching. Teachers usually do not divorce themselves from what they believe to be essential for student learning. For Black teachers, like Dr. Wilson, they often incorporate their identities and their experiences—triumphs and struggles—in their curriculum and teaching.

This Teacher

Dr. Wilson is a female African American English teacher who lives in the Stevenson County School District. The names, district, and school have been changed to mask their identities throughout this manuscript. Having earned her doctorate from a large Midwestern institution, she has been teaching in the district for eleven years but has been teaching for twenty-six years total. Dr. Wilson was the only African American teacher at Stevenson High to teach in the academic core. There were two additional African American teachers who taught in the vocational/elective areas at the school. Energetic and passionate, Dr. Wilson keeps her students laughing and "entertained." She enjoys reading, traveling and, most of all, her own two children.

This Research

As an African American faculty researcher and a former high-school English teacher in a predominantly Black high school in the United States, I wanted to study other teachers' processes in developing multicultural curriculum from various contexts. Employing a case-study approach (Stake, 1994), I sought to understand a Black teacher's use of official curriculum and the enactment of such decision making in a predominantly White school, particularly as issues of race and culture emerged. Race, of course, is socially and legally constructed (see, for instance, Ladson-Billings & Tate, 1995; Spring, 2002; Tate, 1997). It does, from my research and experience, involve the interpretation of an individual's skin color; it also involves so much more. It refers to a person's ancestry: his or her genetic makeup. The reference to genetic makeup does not suggest that individuals from different biological groups are more different than alike. It suggests, however, that race is so ingrained in how we view each other, that it cannot be overlooked or ignored (Bell, 1992; Ladson-Billings, 1998). Thus, when teachers adopt color-blind ideologies, they are rejecting realities of the larger society and that of the school (Johnson, 2002; Lewis, 2001). Race is so ingrained in our country's thinking and educational system, that it cannot be ignored—even if we want to pretend that we are all the same and hence experience equals opportunities, privileges, and advantages. It is important to note that Dr. Wilson often thought about and discussed issues of race and culture in complementary ways, not as separate constructs.

Specific and interrelated questions guiding this research included: (a) How does Dr. Wilson transform her curriculum to have gendered, cultural, and racial meaning? (b) How does Dr. Wilson's transformative curriculum look in action? (c) What is the nature of Dr. Wilson's implicit curriculum and what are students learning (or not) from Dr. Wilson's choices? and (d) How does Dr. Wilson perceive her experiences in developing a transformative curriculum as one of the only Black teachers in the mostly White school? In essence, this research considered what a multicultural curriculum really meant in a predominantly White teaching context when a teacher deliberately developed, implemented, and transformed her curriculum and lessons with racial and cultural meaning. Further, this research focused on how the teacher negotiated her own experiences and her students' experiences to maximize learning opportunities.

Over a five-month period, I conducted context observations and interviews with an African American English teacher. An African American teacher was selected because I wanted to understand the African American teacher's thinking about race and culture in a predominantly White context and how she was able to transform her curriculum and teaching in what could have been a difficult context. Throughout the study, I attended and observed the teacher's classes, randomly attended other school-related activities (e.g., a band concert and a school play), and visited other locations, such as the library and the cafeteria. I was typically in the school for the entire day, one or two days per week. While there were weeks when I visited the school two days, the majority of the time, I was there one day. I was always there at least one day per week.

Most mornings, I was in the school before the bell rang, talking to students and teachers, as well as reviewing my field notes or documents, which were recorded or collected respectively throughout the study. For instance, the teacher shared her plan book with me, as well as worksheets, novels, videotapes, and other materials to help me gain a deep understanding and knowledge base relative to her thought processes around decision making for student learning. Although I participated in some of the classroom tasks, I was more of an observer than a participant. In some cases, I participated in group discussions or commented on themes as they emerged in a particular reading. Most of the time, however, I observed the classroom context.

In addition, I conducted five one- to two-hour, semiformal, structured interviews with Dr. Wilson. These interviews typically took place during the teacher's lunch hour, planning block, or after school. Interviews with the participant were tape recorded, and *I* (rather than a hired transcriptionist) transcribed the tapes to gain a deeper level of "intimacy" with the data. Data were hand coded. Essentially, analysis followed a recursive, thematic process; as interviews and observations progressed, I used analytic induction and reasoning to develop thematic categories. Because findings were based on both observations and interviews, the patterns of thematic findings emerged from multiple data sources, resulting in triangulation. For instance, when the teacher repeated a point several times throughout the study, this became what I called a "pattern." When what the teacher articulated during interviews also became evident in her actions or in her students' actions, this resulted in what I called a "triangulational pattern."

This School

For the purposes of this study, I wanted to select a teacher who was race and culture conscious—who intentionally developed, incorporated, and transformed lessons of race and culture in his or her work to think about student learning opportunities. Stevenson High School was an economically affluent, Midwestern, suburban high school. It accommodates approximately 1649 students with a mostly homogeneous group of enrollees. Specifically, 86 percent of Stevenson High students were European American, 4 percent were Black or African American, 10 percent were Asian American, with 2 percent speaking limited English, 2 percent coming from low income homes, 7 percent receiving special education, and a 3 percent turnover rate.

According to a Stevenson county realtor, houses in the district ranged from $150,000 to $300,000. It was one of two high schools in the Stevenson County District and was known for its competitive soccer and lacrosse teams. Constructed in 1992, the school building was brick, and the architecture was moderately new and sophisticated. Carpet lined a large portion of the commons area, and the cafeteria resembled a coffee shop more so than a school cafeteria. Students often congregated here before school began, during lunch, and after school. The hallways were light and airy, and artwork was displayed throughout most of the hallways, including original pieces by Stevenson High seniors.

Know Thyself: A Precursor to Understanding "the Other"

One way Dr. Wilson transformed her curriculum was by providing a set of experiences to guide her students through self-reflections as they worked to understand others. In Banks's (2003) words, "Students need to understand the extent to which their own lives and fates are tightly tied to those of powerless and victimized groups" (p. 81). Dr. Wilson thought deeply about her students as she made decisions for their learning. This point is interesting if we consider the fact that the majority of her students were White, and she interpreted the students' needs as ones central to understanding race and culture. Dr. Wilson believed that she needed the students to gain a level of self-knowledge and awareness as they attempted to understand others. To explain, Dr. Wilson reported,

> They need to know . . . especially in my senior composition courses, I try and go beyond the Euro-centric literature. Most of the contributors to literature were made by White men. The main thing I consider, besides my kids, then, is the importance of exposing the kids to writers who make up the world: the Hispanics and the Hispanic Americans, the Asians and the Asian Americans, the Africans and the African Americans. You see? Women—women are also important because most of the writers were White men. I want my girls to read about women, too. I try to broaden their horizons. They are on their way out into the real world, and everybody they meet in the world might not look like the people here. To me, this is what's important in the decision making.

Here, Dr. Wilson asserts a need to connect her students' own experiences with new ones that may enlighten them about issues of race and culture. She especially wanted the female students to be exposed to women writers. Clearly, as hooks (1994) explained, women are often left out of the conversation and content in the classroom. At the same time, Dr. Wilson wanted the students to explore writers from other ethnic backgrounds. Thus, issues of culture (if we think about gender as a cultural dimension) were central in Dr. Wilson's thinking about student learning. Exposing the students to "the other" and to "the self" was important to her given that,

> many of these kids don't have any idea about how other people around the world live. They are sheltered. They are good kids, but they just don't *see it* [emphasis added]; so because for years I've been the only Black teacher they encountered, I try and plan and develop a set of experiences for my students that will make them better human beings when they leave. That's why we all participate in self-reflections to help find our niche per se. We have to discover what we're good at, you see? And part of it is helping them understand that we aren't all the same, and it's not their fault that things are the way they are, but we certainly need to know about it. Knowledge is power, and power is change.

Dr. Wilson was committed to having her students engage in self-reflections not only to help them find their "niche" but also to help them think about themselves in relation to others. Clearly, Dr. Wilson was attempting to help her students develop the skills, knowledge, attitudes, and dispositions necessary to function well in the world—a point that multicultural theorists consistently point to as essential for all students (Arias & Poynor, 2001; Banks & Banks, 1995). This same commitment to students is evident in the literature about Black teachers. They believed that preparing students with the tools necessary for success in society was their role and responsibility (Foster, 1997; Siddle-Walker, 1996). They did not pass the blame or the responsibility to others; they took on the responsibility themselves. They felt that their curriculum had to do more than provide a set of learning opportunities from a textbook. The teachers saw a need to expand their curricula to meet the needs of their students' lives outside of the school (Mitchell, 1998).

In a number of meaningful ways, Dr. Wilson's push for self-understanding as the students attempted to understand others can be connected to how she thought about herself in relation to her students:

> The kids, Rich, I consider the kids in this decision making. I work hard to make all my students feel like they are a part of the learning environment. *This might be because I haven't been made to feel like I am accepted in this school* . . . I have been hurt here, and I don't want my students to feel hurt for being different—you know—we need to celebrate our differences. So, when they ask me to teach the seniors who[m] no one else wants to teach, I accept. That's right, Rich, I gladly accept because I bring out the best in them. And they see that I am different, and I have come to love myself for being different because wherever we go in life, there will be people who have problems with us . . . What can I do to make them [students] feel accepted and to appreciate themselves for being who they are? It makes me feel better, too, because they remind me that I am okay. This is especially true when they [the students] ask me to present them with their diplomas at graduation.

Dr. Wilson perceives herself as "different" in this context, mainly because she is one of three African American teachers in the school. In addition, she has felt ostracized because she is quite vocal about issues of racial and cultural diversity in this context, which has made some of her colleagues "uncomfortable." Thus, Dr. Wilson draws on her own personal experiences through introspection in the school and uses these experiences to connect to how her students must feel as "seniors who no one else wants to teach."

Clearly, Dr. Wilson discusses her thinking about decision making in terms of balancing the "self" with that of the "other." In a sense, she is stressing what West (1993) asserts when he suggests that we cannot work for freedom [or social justice] on behalf of others until we are free ourselves. The balance of self and other manifests itself in how Dr. Wilson develops lessons, how she thinks about those lessons, the types of activities she develops to carry out those lessons, and how she tries to help her students maintain or understand their multiple identities. These multiple identities are central in understanding the students' sense of "belonging" and "acceptance" in the larger community of students.

There is an interesting level of transformation in which Dr. Wilson engages. This level of transformation is cognitive in nature. That is, she is working hard to help the students think about multicultural content and to broaden and/or change their mindset in order to think about others in more relevant and meaningful ways. She pushes her students to think about their own role in the school in relation to others in the school in order to help them gain knowledge as they think about others outside of the school. In her words,

> These students will be stronger if they find a sense of belonging. They know that Dr. Wilson cares about them, and I'm not willing to let them get away with mediocrity. Just because they are not all on the football team or the most popular kids, or the kids on the honor roll, does not mean that they are not good students. I know how they must feel. I plan assignments like the self-portrait that allows them to think about themselves in positive ways. I also do self-portraits with them so they can see that I too am different and yet I'm okay. Kids like this, and they do it because they can relate. This helps me every time I do it [write a self portrait], so it probably helps them as well . . . [This] allow[s] them [the students] to see that they too are special. I plan in a way that I can help the kids find worth and see that we don't need to hurt other people because we are different. You know?—it's like, what can I do to link these points subtly in the lessons?

As evident in this passage, Dr. Wilson consciously had students think about themselves through the activities in which she asked the students to engage. Moreover, she also participated in these activities with the students. Pedagogically and philosophically, then, Dr. Wilson was willing to negotiate her power, and she did not consider herself the only, nor the main, arbiter of knowledge. She was transforming the curriculum in a way that honored the students' experiences and thoughts in conjunction with her own. The content of her activities could have easily taken on a more general approach or even focused on other topics, but she was deliberate in her goal to help the students make connections to and draw from their individual experiences of marginalization within the school: they were not "all on the football team" or the popular students; they were the students "who no one else wanted to teach." Dr. Wilson helped the students think about themselves and their own marginalization in order to guide the students to places of empathy (not pity) with others. In essence, Dr. Wilson transformed her curriculum in order to help students view issues, perspectives, and others with their *hearts* as well as with their *heads* (Banks, 2003). Dr. Wilson was very adamant about helping her students connect affective and cognitive experiences.

Dr. Wilson thought that her colleagues were not as concerned about culture as she was. She purposely developed experiences that she was convinced would likely be omitted in her students' education without her exposure. She explained,

> If I were an Asian student, I would want my teacher to know something about Asian writers. It is important to make lessons relevant to students . . . these kids know so much nowadays, so you have to work and make the work fit into their scheme of thinking. So, that means that I got to learn about Asians by asking the right kinds of questions because you don't want to get too personal, but you want them to know that you [as the teacher] care about them enough to learn about their culture. Those things are important, and I think my colleagues could learn a lot from me if they would just ask me, so I try to keep that [the need to learn about other cultures] in mind as I'm preparing to teach my kids.

As evident in this passage, Dr. Wilson acts as a leaner within the classroom, particularly among student "cultures" that were unfamiliar to her. However, she does caution that delving into some of these issues could cause "discomfort" for students, so it is important to approach such discovery or research with care. She also believes that collaboration was important and that her colleagues could learn from her just as she could learn from them. In a sense, Dr. Wilson believed that it takes a village to educate students. Further, based on this evidence, in order for teachers to effectively and realistically reach a level of curriculum transformation, they must invite the students into the process of transformation. The teacher cannot expose students to the wealth of knowledge available about the students in their classrooms, and people outside of the classroom, without relying on and learning from their students. Effective curriculum transformers are at the heart of authentic learning. Both teachers and students participate in the development of knowledge that is coconstructed, cobalanced, and conegotiated (Foster, 1997; Gay, 2000; hooks, 1994; Ladson-Billings, 1994).

The link between the transformative and social-action curriculum of Dr. Wilson emerges in the way that she talks about her work and her belief systems about what students needed:

> I love my kids—all of my kids. You know that. I give out the most diplomas at graduation, and the kids know that I love them, too. What I am saying, however, is that I want to teach my kids more than what's in a book. I want to teach them about hate and how hate is relegated to all groups of people including the White race. I often discuss with my students how the Irish were mistreated. The Italians, you know, they were lynched and could only work in farming and agriculture. I want my kids to eradicate hate. I want them to love themselves because when we hate ourselves, we hate others. Knowledge is key . . . [So], when I plan I think about ways I can reach the kids and help them love the greatness of their people and themselves as individuals so we might promote love for others . . . and then in class I make it plain what's happening. I don't try and beat around the bush.

The example of Dr. Wilson's thinking about her decisions exemplifies the relationship between the transformative and social-action approaches that Banks (1998) described. While on the one hand, Dr. Wilson is working to transform the students' thinking by exposing them to other cultural and racial knowledge, she is, on the other hand, also attempting to have the students change their actions or behaviors by "eradicating hate" and by "loving themselves." She wants her students to learn about "more than what's in a book" and to take charge by transforming their actions in a quest for social justice. Interestingly, she does not provide a set of experiences (or action-oriented tasks) that may more closely reach the social-action stage. Dr. Wilson seemed to adopt and employ a philosophy that once the students became more knowledgeable and conscious about some of the social ills and injustices inherent in our world, the students would work to end the causes and perpetuators of them. Her links to social action, though, were focused on the students' thinking and belief systems and not necessarily on specific activities or strategies that the students might use to address inequity, injustice, and inequality. Indeed, thinking about Dr. Wilson's pedagogical and curricula decision making has significant implications for Dr. Wilson's implicit curriculum as she works to provide learning opportunities that are "more than what's in a book."

Intentional Implicitness

Several researchers and theoreticians discuss the implicit or hidden curriculum as student learning opportunities not as those that are not stated, overt, or explicit in curriculum documents, textbooks, state standards and the like but rather as opportunities to learn that are available to students that are intended or unintended (Anyon, 1980; Apple & King, 1990; Eisner, 1994; Hale, 2001; McCutcheon, 2002; Schubert, 1986). These learning opportunities emerge through the curriculum that the teacher or school provides but are not deliberate or intentional. For instance, a teacher's goal to have students pick up paper between classes might be thought about as an implicit learning opportunity—students are learning from a curriculum, but it is not necessarily written. There were multiple learning opportunities available for student learning that were inherent in Dr. Wilson's implicit curriculum and that were also transformative. They were, indeed, intentional, but not inherent, in the explicit curriculum. For instance, Dr. Wilson explained, "I keep my hair cut short because I want my kids to see a dark skinned Black woman with short hair. For many of them, Rich, they have never seen this. And yes, there are lessons in that. And, yes, I plan this. I am aware of what I'm doing. I want them to see a Black woman with Black features and how yes, I am okay, and I am smart, a reader, successful, you see? I tell them about how I've traveled the world, and they look at me in awe."

She continued teaching me, explaining,

> Their interactions with me and their acceptance of me will help them to take the time to be accepting of other people who may not look like them or have the same kinds of experiences. So, I think culture plays a role. Because if you think about it, why haven't we, in our educational process, taught certain things in the classroom? I mean, if you look at White teachers—how many of them know about Toni Morrison? How many of them know about Terry McMillan? How many of them know about bell hooks? And they are not teaching that kind of stuff. bell hooks is a literary genius; she's more intellectual than many other writers, but my White colleagues teach what they know. They teach Hawthorne, you know, and it's a cultural thing. But with me, you know, we [Black teachers] have to know Hawthorne, too, and that makes me more knowledgeable and flexible because we know both worlds. And I tell my students that people have stories that are not a part of the White story. So, I think culture is so important. And so, these students miss out because White teachers aren't reading these stories.

Several points warrant consideration here. For one, Dr. Wilson points to her perception that we teach what we know—suggesting that the worlds we know, live in, and negotiate influence how we conceive and represent matters inside and beyond the classroom. She discusses her knowledge about multiple roles—suggesting that she "knows" two worlds or realities.

This notion of knowing two worlds points to what Du Bois (1903) refers to as "double-consciousness." Double consciousness suggests that Black Americans often operate through multiple lenses and realities as they experience the world. They are African and they are also American, and these levels of consciousness are often at odds, but they are often in communion as well. Moreover, as evident in the literature, Black teachers often prepared their students for their current reality but also thought about what students would experience later in life (such as after integration; Milner & Howard, 2004). Dr. Wilson seems to be quite aware of both worlds: "Euro-centric literature—It's expected. But for White teachers, they only have to know their own culture, and they don't step outside of that. It's like Black teachers have to know more—we have to know about what's accepted and expected from a European perspective, and we are expected to be the expert on everything Black too. It's hard work."

Here, Dr. Wilson expressed her view on how she is expected to know more than her colleagues. She believed that White teachers tend not to learn material outside of their comfort zones, mainly because they do not have to or because they are not expected to. She further explained this point:

> Whites read about things that interest them, and many of those Black writers are not of interest to them. You know, but when you're minority, and I hate using that word—minority, it's such a negative word—but I'm going to use it for the sake of this interview and your notes that you need to take. When you're a minority, you have to read and know all about the White writers. I remember how I felt when I did my master's. We were expected to intimately know about all the White writers. It's like Mark Twain. I try and understand why he felt he could imitate the Black dialect. They believe that we are not complex. It's simple to understand us, you see? But as a minority, we are expected to intently study White's history because we are not always expected to succeed. This was the case when I did my doctorate. You know how it is.

In addition, the notion of respect can be considered as part of Dr. Wilson's implicit curriculum. That is, lessons of respect were not central to the explicit curriculum, but were certainly consistent with Dr. Wilson's implicit curriculum. In her view, "many of the teachers here don't deal with these issues. It's not because they are insensitive to it, but because they don't live it. You know, Rich, we [referring to herself and me] live it when we walk down the street or go into a grocery store, and I tell my [own] kids [her biological children] that we are still dealing with this hate. So, if one or two of them leave[s] and [has] a different view on this, I am happy."

Dr. Wilson's implicit curriculum did not stop with respect and keeping her hair cut short. She also provided some nontraditional learning opportunities for her students outside of the school:

> I want the students to know me outside of school, and I want them to know who I am culturally, you know, how I live from day to day, who my [own] children are, what I have in my home. You know I have African art throughout my home. I lived in Africa for years with my ex-husband, and the kid'll ask, "Dr. Wilson, where did you get this?" What does it symbolize, and those are a part of the teachable moments I keep talking about. For most of them, they have never been inside of a Black person's house, you know?

In short, Dr. Wilson has invited her students into her home. These "teachable moments" were frowned upon by many of Dr. Wilson's colleagues: "Why would I want them to come to my home? Why would I want them to call me?" As Dr. Wilson explained, "I don't take this attitude. I do this because years ago when I was a student, my teacher went to my church and would come to my home, you know? Sunday dinners—my teachers would tell my mom things that [Dr. Wilson] was doing or not doing. It was that kind of relationship that has brought me to the point where I am. It's probably a cultural thing. I believe that these are teachable opportunities, and I take advantage of them." Thus, as evident in the literature about Black teachers, Black teachers were inquiry minded and attempted to learn about their students' outside worlds by visiting the students' out-of-school contexts, such as grocery stores, churches, and/or their homes. As evident from this research, Dr. Wilson is allowing her students to learn about her own experiences, and she points to the fact that her teachers used to visit her and her church.

An important connection to Dr. Wilson's thinking about diversity and her teaching centered on her past (and more established) and present experiences. She talked about her childhood experiences around race and culture, and she also talked about her more current experiences around them. Brand and Glasson's (2004) important study of preservice science teachers revealed the profound impact on "how early life experiences and racial and ethnic identities of . . . teachers influenced both their beliefs about diversity in science classrooms and science teaching pedagogy" (p. 119). In addition, the study pointed out how the teachers' "self-images" (p. 138) shaped their curriculum decisions and teaching. Accordingly, Dr. Wilson's early experiences shaped what she came to believe about what and how students ought to learn, and those past experiences continued to emerge throughout her life and consequently also showed up in her teaching.

Clearly, Dr. Wilson's implicit curriculum of allowing students to visit her in her home was a practice that she linked to her own schooling and early experiences. This point connects to what I discovered (see Milner & Howard, 2004) about Black teachers and their teaching prior to the *Brown v. Board of Education* (1954) decision. In essence, the study revealed that Black teachers participated in the community and saw themselves as members of the community, and so students often experienced learning from them in multiple sites (such as in church or even in their home).

While Dr. Wilson planned and implemented lessons that considered race and culture, she also informed me that such foci had caused her problems with other teachers and her administration. Those problems had been most prevalent during her early years of teaching at the school, but they still affected her.

The Price Paid for Decisions

Clearly, Dr. Wilson wanted her students not only to better understand differences and similarities among people but also to take that knowledge and become change agents "through their knowledge about the social and political" nature as exposed through literature, class discussions, and their personal experiences. Her desire to have her students become change agents is right in line with Banks's (1998) social-action approach. However, again, she does not necessarily facilitate any actual activities for the students. Her goals for social action are embedded in her desire to change their thinking about "assumptions" and actions with others. As she noted,

> I've become aware of how many of my colleagues look at various selections, and I'm not sure if they understand the whole message of what writers are attempting to do. What I want to talk about is basically how we teach a book called *To Kill a Mockingbird*—how we don't intently focus on the political, social, and racial aspect of that book. There are two parts to that book. There is Boo Radley, and then there is that Tom Robinson situation. When I first started teaching here and I taught that book, my students would run down to the principal's office and tell the principal that I was prejudiced—they thought I was a racist. Maybe I did spend too much time on the Tom Robinson aspect, but that's how they handled our discussions. They thought I was racist for exposing areas that made them uncomfortable. I was trying to get them to see how the female character falsely accused Tom Robinson, [a Black male character], of raping that woman [a White character]. And on the other side, I was telling them that there are two parts of that book, because the other part is how we make assumptions about people like Boo Radley. You know, Boo never came out of that house—the house was all spooky and so

> on and so forth. And we make assumptions about people . . . my students didn't see all that. And as you know, I'm sort of animated in my classes, and they misinterpreted what was going on in there. I didn't have any Black students in that class my first year, and I've found that when I do have Black students or Asian students or Hispanic students, they see what I'm trying to do and say because we live those experiences. It is my White students who don't understand, and I say that to say that there are cultural differences that exist when you teach in a predominantly White situation. They [cultural differences] ought to be explored if we are doing what we need to do—preparing our kids for the real world.

In this quote, a few issues are expressed. Dr. Wilson makes it explicit that there "are cultural differences that exist" in predominantly White teaching contexts although many of her colleagues did not think so. This point is also stressed in Lewis's (2001) research about a predominantly White elementary school. The White adults in the school basically did not conceive of race as important in their decision making because there were few students and teachers of color in the school and the community. However, through Lewis's careful analysis, she learns that race is central to the function of the school, interactions between and among students, and how adults (parents and teachers alike) thought about pertinent issues. In addition, Dr. Wilson's wanting her students to act differently was framed by how she wanted them to think about certain issues in novels, such as *To Kill a Mocking Bird*.

Dr. Wilson wanted to find the social and political themes in the readings and to use them to provide learning opportunities that would cause her students to act differently—perhaps in line with a social-justice agenda. She planned and implemented lessons that focused on social and political matters in her classes and had been doing so for many years. This was done, however, at the risk of being accused of being prejudiced and a racist, as "many of the students took my teaching in this way as an attack on them. And, of course, there is a price we pay when you go outside of people's comfort areas. I was called into the office [by the principal] and questioned about it. And you guessed it—eventually my principal took those freshmen away from me. So, there is a price you pay for decision making and teaching in this way."

Attempting to transform the curriculum in such a profound manner certainly had its consequences. Dr. Wilson's teaching in a predominantly White context with colleagues who rarely thought about such issues likely intensified the possibility of her being ridiculed. Thus, helping teachers develop the skills to transform the curriculum in ways that exemplify racial and cultural awareness is not enough. *Teachers must be prepared to negotiate, balance and to combat pervasive discourses and practices that already exist in a context* (Buendia et al., 2003; Milner & Woolfolk, 2003; Siddle-Walker, 2000). Dr. Wilson elaborated on the resistance she perceived among her colleagues: "I think that it's due to the privileges of the White population here. Whether they be children or adults, teachers, my co-workers, there are privileges that they have been receiving that many of these kids don't understand." As Gordon (1990) explained, "Critiquing your own assumptions about the world—especially if you believe the world works for you" (p. 88) is indeed an arduous and complex endeavor.

Accordingly, her thinking, decision making, and teaching were disputed by many of the students in that class and, as a result, "that class was taken from me. These were ninth graders, and I don't teach ninth graders anymore, haven't for years now because they felt like many of the students were right—that I was a racist. And that hurt me, Rich, when I first came to this building because I wasn't a racist. I am not racist. I just wanted to *open eyes* [emphasis added]. I was anxious for the students to *see it* [emphasis added]." This has not, however, stopped her from working to eradicate "ignorance where culture and race is concerned," as she expressed these same issues to her seniors. As she explained, "The older kids see it much clearer than the younger ones." In these senior courses, again, she explicitly stressed issues that were important to her through her experiences as a person of color. In her words,

> So, they get a taste of it, and I can openly tell them that I want them to not go away to college not having experienced other people's cultures. But I found that a lot of my ninth grade students, it was like they were in denial, and I think that it had to do with maturation (sic). But what surprised me was like not so much the kids, it was my administration that didn't understand because I was constantly being called down to the office and told that the students said that I was a racist for teaching certain books in certain ways . . . and now it is more acceptable to teach certain books like *To Kill a Mockingbird* because White teachers are now teaching that book. You see, they see the worth so it's accepted and even expected now. *But* they [White teachers] are not teaching the social, political, and historical aspects of that book . . . Students can really learn from that book because it has implications for their lives today. It's like a classic, and classics have themes that relate to our lives . . . and racism is a major theme that has been blatantly overlooked by most teachers here—how we treat people . . . They are exposed to those parts of the book when I taught it, you know?

As noted here, there were consequences to what Dr. Wilson decided to teach (the curriculum) and how she taught it (the instruction) when she first entered the district and school. She exhibited a level of persistence that many teachers likely would have found unbearable. The consequences of such interaction and perceptions of colleagues and administration could cause a teacher to revert to or to fall in line with the common discourses established in a school (see Buendia et al., 2003, for more on this). However, those experiences, although hurtful, did not stop Dr. Wilson from working toward "eye-opening" experiences when considering the cultural and racial differences that developed and linked to difference, assumptions, and, ultimately social and political awareness. She has to incorporate these issues with her seniors now rather than with the freshmen—but they are still integral to her curriculum and teaching.

In addition, Dr. Wilson stressed that her colleagues were not teaching certain books and themes in "appropriate" and transformative ways. In Dr. Wilson's perception, her colleagues adopted an additive approach (Banks, 1998) in the development and enactment of their curriculum (particularly novels and themes) with cultural and racial meaning. In her view, the teachers were teaching at a lower level because they were not deeply transforming the curriculum. They were adding a book here and there without infusing and transforming the entire curriculum in profound ways.

Transformative Thinking and Decision Making Enacted

Transformative teaching, for Dr. Wilson, was done in some nontraditional ways. For instance, as already discussed, she invited her students to her home and often talked to them on the telephone before and after school. She expanded on how her thinking and decision making transferred into her teaching, stating,

> Kids are into drugs; kids are, you know, smoking marijuana; they're not always being quiet when you tell them to be quiet. Homes are different. We have what they call a "blended family." We have divorced families. So, all of those things are important in kids' lives. My kids know that I was married, and now, I'm divorced, and my life goes on, and it's okay, you see? So, I guess part of it is showing the differences, but part of it is showing the similarities. We all hurt. We all cry. We all have bad days. I want them to see me as a person who is different and regardless of how we try and portray this "we are the world" attitude; I am still different and treated differently because of the color of my skin and because of my cultural heritage. And then, when I show them the differences, I show them just how similar we are, too. So, being different is not bad 'cause we are actually a lot alike. Does that make sense?

The way Dr. Wilson invites her students to her home and talks to her students on the phone before and after school has a profound impact on how she is able to connect with her students, to learn from them, and to help the students think about both differences and similarities at the same time or in complementary ways. Dr. Wilson's transformative pedagogy actually meant that she helped her students think about differences and similarities.

A Representative Class Session in Dr. Wilson's Senior Composition Class

Dr. Wilson prepares the class for the day's reading—Alice Walker's *Everyday Use*, a short story that focuses on the cultural and historical legacy embedded in patch quilts and the art of quilting. She provides some brief information about Alice Walker and the story. She calls this "setting the stage." "You see," she informs, "Alice Walker is a brilliant African American woman who wrote the novel *The Color Purple*, which was made into a movie. Has anyone in this class read that novel or seen the movie?" Not more than two students raise their hands (all of the students were White or European American in the class). "I see," Dr. Wilson responds. "Well, let me tell you a bit about that novel before we move on." She spends about ten minutes discussing the time period of the novel and points out some of the major themes. After this, she returns to the story at hand—*Everyday Use*.

The students read the story with Dr. Wilson orchestrating the reading: she calls on students, one by one, to read—there are no choices in the matter—"you either read or you are not participating, which means points are deducted." After a few paragraphs, Dr. Wilson interjects, "I remember when I was a girl. My grandmother made me a quilt when I went off to college. I still have it today. That quilt is old and dingy now, but it'll be in my family forever because it means something to me—culturally. I see that quilt as more than pieces of cloth sewn together." Dr. Wilson continued, "And I miss my grandmother—the food she cooked, and her smell, you know, all the things that remind us of the good times—when my sisters and I would play in her yard, and the clothes on the line outside flew in the wind. Those were some great times. And, so, when I run across that quilt in my basement, I think back to those times and I get full [emotional]. Those are times for me to share about my sisters, my parents, and my grandmother with my [own] kids."

Not a person moves during Dr. Wilson's soliloquy. Engagement—words cannot describe the solace in the room. Dr. Wilson continues, "And if my grandmother . . . could see me now as *Doctor Wilson*, their [her grandmother and other family members'] hearts would be glad. I was a little Black girl who grew up right outside of [Ohio], and so, yes, it is deeper than just the cloth, and the quilt, and the fabric. It's about artifacts that we treasure and allow to become meaning makers to remind us—to remind us to tell stories to our children who I hope will tell stories to theirs—starting from a patch quilt. I cannot help but think about this as we read this story." The bell sounds to dismiss class, and no one moves. It is obvious that this is a special moment not only for Dr. Wilson but also for her students.

As evident in the above description, Dr. Wilson relied on family customs and traditions as she enacted her lessons and developed and enacted transformative curricula and pedagogy. This was consistently the case as I observed her classes. Clearly, there is a reliance on her family's customs, traditions, and her multiple identities and experiences of quilting that Dr. Wilson reflects on and implements in this lesson. The transformative curriculum and pedagogy have at their core a sharing of the self (as teacher) to help guide students toward self-thinking, reflection, and understanding.

Later, during an interview, I asked Dr. Wilson about this class session. In her words, "You know, Rich, I don't mind sharing my experiences with them. Some of them [her experiences] are hurtful, and others are times to celebrate. They see me for a real person because I cannot present myself in any other way. They know this about me, and I am proud of that." She went on to describe a student in that class who appeared, through my observations, captivated by the lesson during this session. Dr. Wilson explained,

> During another class session, Dan talked about being upset about the people who are coming into our district. I think he was referring to the Spanish or the Mexicans; I'm not sure what nationality. And Dan was saying that he was upset that this guy couldn't understand him in a convenience store. He said, "Why are these people moving here, and they can't even speak English?" And that gave me a chance to teach—to really get down and teach. I love it—because this kid was very passionate when asking, "Why are these people here?" And I told him that these people are doing things that we don't want to do. He was saying that he didn't like these people coming here, and I was saying, that his family, his grandparents, great-grandparents came from different countries. And they don't see that; that's all important to me because I think that sensitivity, Rich, I have a sensitivity to people who are not like my kids at Stevenson High School—that are not, you know, mainstreamed right away.

Cleverly, she then links her comments to the power structure: "Anytime you're part of the ruling class with power, it's likely that you don't see it. You're not as sensitive to minority groups, and that's okay. We have to learn these things from people who live them. And Rich, it's not just Black and White. I've had many of my Asian students talk about things that they've heard or gone through, too. We all learned from the discussion that day, even Dan." Similarly, with respect to power and privilege, Delpit (1995) wrote, "Those with power are frequently least aware of—or least willing to acknowledge—its existence. Those with less power are often most aware of its existence" (p. 24).

The students often struggled to see Dr. Wilson's points and to realize and acknowledge their own power. Although Dr. Wilson has some power as the teacher in the classroom, she is quick to gauge the power and privilege her students have by virtue of their race (McIntosh, 1990; Milner, 2003b). Dr. Wilson is able to discern her students' racial capital because of the injustice, oppression, and marginalization she has experienced herself.

Students' Acceptance of Dr. Wilson's Curriculum and Teaching

The students in this study appeared to embrace Dr. Wilson's curriculum and teaching. This was made evident by the nature of their participation and engagement in the classroom, their interactions with her in the hallways, and their invitations asking her to award them their diplomas at graduation. Dr. Wilson prided herself on the fact that she awarded more diplomas to graduating seniors than did any of her colleagues. Because the students selected their most influential teacher to award their diploma, Dr. Wilson interpreted the students' invitations as a sign that she was making a positive impact on them and their experiences.

A Teacher's Transformative Curricula and Teaching

Dr. Wilson's transformative multicultural curriculum seemed to be effective in the predominantly White setting. To achieve this, Dr. Wilson helped the students think about themselves and how privilege, power, and marginalization connected to their experience. In short, Dr. Wilson helped the students understand their economic and ethnic privilege while, simultaneously, facilitating her students' reflections on their own experiences of marginalization (most of the students were not on the football team or the most popular students, in Dr. Wilson's view). Transforming the curriculum also meant that students needed exposure to a curriculum about others—be it other cultures or gender groups. Another area of curriculum transformation was that of relational understanding. That is, Dr. Wilson attempted to help her students understand *themselves in relation to others*. As evident from this study, effective transformative curricula must be designed to help students focus on the self (whoever the students may be) and the other.

While it was clear that Dr. Wilson wanted her students to gain a deep understanding of certain aspects of multicultural education and to act on their understandings, I never observed, recognized, or heard anything that amounted to action-oriented decisions. Dr. Wilson's focus was mainly cognitive—helping the students *think about* ways to make a difference and embrace social justice—abstractions such as "eradicating hate" and "celebrating differences."

In addition, the support (or lack thereof) that Dr. Wilson received from her colleagues, the community, and the school culture seemed to have a profound influence on her desire to develop curricula and instructional techniques that helped her students develop multicultural consciousness and competence. Some teachers would have wanted to retreat from such a focus without support from their colleagues. Similarly, Buendia et al.'s (2003) research about English-language learners (ELLs) and teaching highlighted tensions between established discourses in a context and the pedagogical, as well as philosophical, beliefs teachers of color bring into the school. In short, the discourse in the school context carried "deficit views of immigrant students" (p. 315). For instance, teachers believed that the immigrant students needed to be "socialized . . . into things like being responsible . . . [and] how to handle all of the freedom" (p. 302) in the new context. The ingrained discourse of the school seemed to convey the message that the students did not bring intellectual and social assets and capital to the learning environment. Moreover, the teachers of color in the study struggled to develop and implement curricular and pedagogical strategies that, from their experiences, could more appropriately meet the needs of ELLs—in opposition to the permeating dominant discourse in the school. As Banks and Banks (1995) explained, "The school culture and social structure are powerful determinants of how students [and teachers] learn to perceive themselves. These factors influence the social interactions that take place between students and teachers and among students, both inside and outside the classroom" (p. 153). Clearly, the environment and the persistent discourses that characterize a teacher's workplace are essential to both what and how a teacher practices.

Although teachers of color may find racial and cultural conditions and experiences appropriate and relevant because of their personal experiences of racism and sexism, for instance, the pervasive (and White) belief systems, goals, missions, and discourses of the school can circumvent highly capable teachers' desires to transform the curriculum. In Buendia et al.'s (2003) words, "The present-day contexts of schools may push critically minded teachers of color in ways that undermine their desires" (p. 317). As is the case in this study, Dr. Wilson perceived that many of her colleagues were resistant to racial and cultural awareness; yet, she continued in her moral quest to provide her students with a set of opportunities that may make them more culturally and racially aware for the real world. As explained previously, she paid a price for this persistence. Teachers (of any ethnic background) may struggle to persevere in such a context, particularly novices. And whether the context was supportive of Dr. Wilson's curricula and pedagogical strategies is not as important as Dr. Wilson's *perception of* her colleagues' support or lack of support. Still, Dr. Wilson had earned a doctorate, having read and been exposed to a wide range of information that likely contributed to her persistence as she faced ridicule. What are the consequences for teachers who cannot rely on such a wide body of knowledge and experience in their quest to disrupt and counter established discourses and beliefs about multicultural curriculum and its development?

As evident in this research, who teachers are as racial and cultural beings often emerge in their curricula selections and implementation. What and how a teacher teaches reflects how that teacher perceives himself or herself as well as who and what a teacher stands for. Moreover, we know that who teachers are, their experiences, and their stories often find themselves in their work with students. Thus, teaching, on certain level, is almost already a personal and political endeavor, and helping teachers understand themselves (their beliefs, values, and philosophies) will make them more effective with their diverse students.

Dr. Wilson, a Black teacher, was quite effective in developing and implementing a multicultural curriculum with her predominantly White students. As Gay (2000) asserted, "Similar ethnicity between students and teachers may be potentially beneficial, but it is not a guarantee of pedagogical effectiveness" (p. 205). In other words, teachers from any ethnic, cultural, or racial background can be successful with any group of students when the teachers have the appropriate knowledge, attitudes, dispositions, and beliefs about teaching, learning, and their students. As evident in this study as well as in the research of Gay (2000), "Members of one ethnic group [can be] successful teachers of students from other ethnic and racial groups" (p. 205) even when the teacher is Black and is teaching predominantly White students about multiculturalism.

Dr. Wilson felt a professional and moral obligation to prepare her students for a diverse world and to help students not be color-blind or in denial about our diverse society and world. Similar to other Black teachers and their moral decision making and dilemmas, Dr. Wilson understood that her job and responsibility extended far beyond the explicitly stated curriculum and other mandates (see Agee, 2004, for more on this). Teaching and learning are preparation for life. More importantly, multicultural education is important in all contexts—even mostly White ones. Dr. Wilson is able to develop a transformative curriculum that will likely ensure that more of her students will work toward social justice, understand and rally *with* oppressed groups, and pass along healthy attitudes to their own children. Of course, additional studies in this area will only complement and enhance our knowledge around these issues. In particular, we need to know more about how Black teachers negotiate, develop, and implement the curriculum as well as persevere in predominantly White contexts. How do these teachers negotiate their own identities and beliefs in order to meet the needs of students who live in very different worlds? This type of knowledge can help us understand and support such teachers. Teacher-education programs must broaden their aims and agendas to include Black teachers; studying these teachers' experiences in a variety of contexts will only assist us in our endeavors to effectively redevelop the curriculum in teacher-education programs.

Finally, as Banks (2003) explained, "The world's greatest problems do not result from people being unable to read and write. They result from people in the world—from different cultures, races, religious and nations—being unable to get along and to work together to solve the world's intractable problems such as global warming, the AIDS epidemic, poverty, racism, sexism, and war" (p. 18). Surely, all students, not just students of color, need and deserve learning experiences and curricula that ensure they acquire the knowledge, skills, awareness, understanding, and dispositions necessary to combat these social problems and experience the world more fully.

References

Agee, J. (2004). Negotiating a teaching identity: An African American teacher's struggle to teach in test-driven contexts. *Teachers College Record, 106*(4), 747–774.

Anyon, J. (1980). Social class and the hidden curriculum of work. *Journal of Education, 162*(1), 366–391.

Apple, M. W., & King, N. (1990). *Economics and control in everyday school life.* In M. W. Apple (Ed.), *Ideology and curriculum* (pp. 43–60). New York: Routledge.

Arias, M. B., & Poynor, L. (2001). A good start: A progressive, transactional approach to diversity in pre-service teacher education. *Bilingual Research Journal, 25*(4), 417–434.

Banks, J. A. (1995). Multicultural education and curriculum transformation. *The Journal of Negro Education, 64*(4), 390–400.

Banks, J. A. (1998). Curriculum transformation. In J. A. Banks (Ed.), *An introduction to multicultural education* (2nd ed., pp. 21–34). Boston: Allyn and Bacon.

Banks, J. A. (2001). Citizenship education and diversity: implications for teacher education. *Journal of Teacher Education, 52*(1), 5–16.

Banks, J. A. (2003). Teaching literacy for social justice and global citizenship. *Language Arts, 81(*1),18–19.

Banks, C. A. & Banks, J. A. (1995). Equity pedagogy: An essential component of multicultural education. *Theory into Practice, 34*, 152–158.

Banks, J. A., Cookson, P., & Gay, G. (2001). Diversity within unity: Essential principles for teaching and learning in a multicultural society. *Phi Delta Kappan, 83*(3), 196–203.

Baszile, D. T. (2003). Who does she think she is? Growing up nationalist and ending up teaching race in White space. *Journal of Curriculum Theorizing, 19*(3), 25–37.

Bell, D. (1992). *Faces at the bottom of the well: The permanence of racism*. New York: Basic Books.

Buendia, E., Gitlin, A., & Doumbia, F. (2003). Working the pedagogical borderlands: An African critical pedagogue teaching within an ESL context. *Curriculum Inquiry 33*(3), 291–320.

Brand, B. R., & Glasson, G. E. (2004). Crossing cultural borders into science teaching: Early life experiences, racial and ethnic identities, and beliefs about diversity. *Journal of Research in Science Teaching, 41*(2), 119–41.

Collins, P. H. (1991). *Black feminist thought: Knowledge, conscious, and the politics of empowerment: Perspectives on gender* (Vol. 2). New York: Routledge.

Cooper, P. M. (2003). Effective White teachers of Black children: Teaching within a community. *Journal of Teacher Education, 54*(5), 413–427.

Delpit, L. (1995). *Other people's children: Cultural conflict in the classroom*. New York: The New Press.

Dillard, C. B. (1996). Engaging pedagogy: Writing and reflection in multicultural teacher education. *Teacher Education, 8*(1), 13–21.

Du Bois, W. E. B. (1903). *The souls of Black folks*. New York: Fawcett.

Eisner, E. W. (1994). *The educational imagination: On the design and evaluation of school programs*. New York: MacMillan College Publishing Company.

Ford, D. Y. (1996). *Reversing underachievement among gifted Black students: Promising practices and programs*. New York: Teachers College Press.

Foster, M. (1990). The politics of race: Through the eyes of African-American teachers. *Journal of Education, 172*, 123–141.

Foster, M. (1995). African-American teachers and culturally relevant pedagogy. In J. Banks (Ed.), *Handbook of research on multicultural education* (pp. 570–587). New York: Simon & Schuster.

Foster, M. (1997). *Black teachers on teaching*. New York: The New Press.

Freire, P. (1998). *Pedagogy of the oppressed*. New York: Continuum.

Gay, G. (1994). Coming of age ethnically: Teaching young adolescents of color. *Theory into Practice, 33*(3), 149–155.

Gay, G. (2000). *Culturally, responsive teaching: Theory, research, & practice*. New York: Teachers College Press.

Gordon, B. M. (1990). The necessity of African-American epistemology for educational theory and practice. *Journal of Education, 172*(3), 88–106.

Grant, C. A., & Tate, W. F. (1995). Multicultural education through the lens of the multicultural education research literature. In J. Banks & C. Banks (Eds.), *Handbook of research on multicultural education* (pp. 145–166). New York: Macmillan.

Grillo, B. A. (1998). Multicultural education: A developmental process. *Montessori Life, 10*(2),19–21.

Hale, J. E. (2001). *Learning while Black: Creating educational excellence for African American children*. Baltimore, MD: Johns Hopkins University Press.

Holmes, B. J. (1990). New strategies are needed to produce minority teachers (Guest Commentary). In A. Dorman (Ed.), *Recruiting and retaining minority teachers*. Policy Brief No. 8. Oak Brook, IL: North Central Regional Educational Laboratory.

hooks, b. (1994). *Teaching to transgress: Education as the practice of freedom*. New York: Routledge.

Hudson, M. J., & Holmes, B. J. (1994). Missing teachers, impaired communities: The unanticipated consequences of Brown v. Board of Education on the African American teaching force at the precollegiate level. *The Journal of Negro Education, 63*, 388–393.

Irvine, R. W., & Irvine, J. J. (1983). The impact of the desegregation process on the education of Black students: Key variables. *The Journal of Negro Education, 52*, 410–422.

Irvine, J. J. (1998, May 13). Warm demanders. *Education Week, 17*(35), 56.

Irvine, J. J. (2003). *Because of the kids: Seeing with a cultural eye*. New York: Teachers College Press.

Jenks, C., Lee, J. O., & Kanpol, B. (2001). Approaches to multicultural education in preservice teacher education: Philosophical frameworks and models for teaching. *The Urban Review, 33*(2), 87–105.

Johnson, L. (2002). "My eyes have been opened": White teachers and racial awareness. *Journal of Teacher Education, 53*(2), pp. 153–167.

King, S. (1993). The limited presence of African-American teachers. *Review of Educational Research, 63*(2), 115–149.

Ladson-Billings, G. (1992a). The multicultural mission: Unity and diversity. *Social Education, 56*(5), 308–311.

Ladson-Billings, G. (1992b). Liberatory consequences of literacy: A case of culturally relevant instruction for African American students. *Journal of Negro Education, 61*(3), 378–391.

Ladson-Billings, G. (1994). *The dreamkeepers: Successful teachers of African-American children*. San Francisco: Jossey-Bass.

Ladson-Billings, G. (1996). Silences as weapons: Challenges of a Black professor teaching White students. *Theory into Practice, 35*, 79–85.

Ladson-Billings, G. (1998). Just what is critical race theory and what's it doing in a nice field like education? *Qualitative Studies in Education, 11*(1), 7–24.

Ladson-Billings, G. (2000). Fighting for our lives: Preparing teachers to teach African American students. *Journal of Teacher Education, 51*(3), 206–214.

Ladson-Billings, G. (2004). New directions in multicultural education: Complexities, boundaries, and critical race theory. In J. A. Banks & C. A. M. Banks (Eds.), *Handbook of research on multicultural education* (2nd ed., pp.50–65). San Francisco, Jossey-Bass.

Ladson-Billings, G., & Tate, B. (1995). Toward a critical race theory of education. *Teachers College Record, 97*(1), 47–67.

Lewis, A. E. (2001). There is no "race" in the schoolyard: Colorblind ideology in an (almost) all White school. *American Educational Research Journal, 38*(4), 781–811.

McAllister, G., & Irvine, J. J. (2000). Cross cultural competency and multicultural teacher education. *Review of Educational Research 70*(1), 3–24.

McCutcheon, G. (2002). *Developing the curriculum: Solo and group deliberation.* Troy, NY: Educators' Press International.

McCutcheon, G., & Milner, H. R. (2002). A contemporary study of teacher planning in a high school English class. *Teachers and Teaching: Theory and Practice, 8*(1), 81–94.

McGowan, J. M. (2000). Multicultural teaching: African-American faculty classroom teaching experiences in predominantly White colleges and universities. *Multicultural Education, 8*(2), 19–22.

McIntosh, P. (1990). White privilege: Unpacking the invisible knapsack. *Independent School, 90*(49), 31–36.

Milner, H. R. (in press). Stability and change in prospective teachers' beliefs and decisions about diversity and learning to teach. *Teaching and Teacher Education.*

Milner, H. R. (2005). Stability and change in prospective teachers' beliefs and decisions about diversity and learning to teach. *Teaching and Teacher Education, 21*(7), 767–786.

Milner, H. R. (2003a). A case study of an African American English teacher's cultural comprehensive knowledge and (self) reflective planning. *Journal of Curriculum and Supervision 18*(2), 175–196.

Milner, H. R. (2003b). Teacher reflection and race in cultural contexts: History, meaning, and methods in teaching. *Theory into Practice 42*(3), 173–180.

Milner, H. R., & Howard, T. C. (2004). Black teachers, Black students, Black communities and *Brown*: Perspectives and insights from experts. *Journal of Negro Education, 73*(3), 285–297.

Milner, H. R., & McCutcheon, G. (2002). A high school social studies teacher's planning and the 21st century: A call for more studies. *Teacher Education and Practice 15*(3), 95–109.

Milner, H. R., & Smithey, M. (2003). How teacher educators created a course curriculum to challenge and enhance preservice teachers' thinking and experience with diversity. *Teaching Education, 14*(3), 293–305.

Milner, H. R., & Woolfolk Hoy, A. (2003). A case study of an African American teacher's

self-efficacy, stereo-type threat, and persistence. *Teaching and Teacher Education, 19*, 263–276.

Mitchell, A. (1998). African-American teachers: Unique roles and universal lessons. *Education and Urban Society, 31*(1), 104–122.

Morris, J. E. (2004). Can anything good come from Nazareth? Race, class, and African American schooling and community in the urban south and Midwest. *American Educational Research Journal, 41*(1), 69–112.

Monroe, C. R., & Obidah, J. E. (2004). The influence of cultural synchronization on a teacher's perceptions of disruption: A case study of an African-American middle-school classroom. *Journal of Teacher Education, 55*(3), 256–268.

Pang, V. O., & Gibson, R. (2001). Concepts of democracy and citizenship: Views of African American teachers. *The Social Studies, 92*(6), 260–266.

Parker, L. (1998). "Race is . . . race ain't": An exploration of the utility of critical race theory in qualitative research in education. *Qualitative Studies in Education, 11*(1), 45–55.

Rushton, S. P. (2004). Using narrative inquiry to understand a student-teacher's practical knowledge while teaching in an inner-city school. *The Urban Review, 36*(1), 61–79.

Schubert, W. H. (1986). *Curriculum: Perspectives, paradigm, and possibility.* New York: Macmillan.

Siddle-Walker, V. (1996). *Their highest potential: An African American school community in the segregated south.* Chapel Hill, NC: University of North Carolina Press.

Siddle-Walker, V. (2000). Valued segregated schools for African American children in the South, 1935–1969: A review of common themes and characteristics. *Review of Educational Research, 70*(3), 253–285.

Sleeter, C., & Grant, C. E. (1994). *Making choices for multicultural education* (2nd ed). New York: Merrill.

Smith, R. W. (2000). The influence of teacher background on the inclusion of multicultural education: A case study of two contrasts. *The Urban Review, 32*(2), 155–176.

Solorzano, D. G. & Yosso, T. J. (2001). From racial stereotyping and deficit discourse toward a critical race theory in teacher education. *Multicultural Education, 9*(1), 2–8.

Spring, J. (2002). *American education* (10th ed.). New York: McGraw-Hill.

Stake, R. E. (1994). Case studies. In N. K. Denzin & Y. S. Lincoln (Eds.), *Handbook of qualitative research* (pp. 236–247). Thousand Oaks, CA: Sage.

Tate, W. F. (1997). Critical race theory and education: History, theory, and implications. In M. Apple (Ed.), *Review of research in education* (Vol. 22, pp. 195–247). Washington, DC: American Educational Research Association.

Tillman, L. C. (2004). (Un)Intended consequences? The impact of *Brown v. Board of Education* decision on the employment status of Black educators. *Education and Urban Society, 36*(3), 280–303.

Troutman, P. L., Jr., Pankratius, W. J., & Gallavan, N. P. (1999). Preservice teachers construct a view on multicultural education: using Banks' levels of integration of ethnic content to measure change. *Action in Teacher Education, 20*(1), 1–14.

West, C. (1993). *Race matters.* Boston: Beacon Press.

Wink, J. (2000). *Critical pedagogy: Notes from the real world* (2nd ed.). New York: Longman.

Part II

Culture, Curriculum, and Identity with Implications for English-Language Learners and Immigration

3

Promoting Educational Equity for a Recent Immigrant Mexican Student in an English-Dominant Classroom

What Does It Take?

Ana Christina DaSilva Iddings, University of Arizona, and Brian Christopher Rose, Vanderbilt University

According to the National Clearinghouse for English Language Acquisition (NCELA, 2006) there are about 5 million students identified as English-language learners (ELLs) currently enrolled in American public schools in grades pre-K through twelve. This number represents about 10 percent of the total enrollment in U.S. public schools. Within the last fifteen years, the numbers of ELLs have increased at a rate almost eight times greater than the total public-school enrollment. It is estimated that about nineteen million ELLs will be attending public schools in the year 2020 (Trueba, 1999). More startling, by the year 2030 the school-aged children of immigrants, many of whom are ELLs, will total nearly twenty-six million (Tienda & Mitchell, 2006). In view of the rapid growth of ELL populations in American schools, in recent years the educational circumstances of these students have been of national concern (see Snow, 2002).

In this chapter we ask the question, "What it would take to promote greater educational equity for recent immigrant children who are ELLs in English-dominant classrooms?" To address this question, we report on a study concerning a research-based professional development project involving a fourth-grade teacher who had little professional preparation in cross-cultural/cross-linguistic teaching, and a recent immigrant Mexican student enrolled in her class. This project aimed to develop deeper understandings about the learning processes of the Mexican student, a native speaker of Spanish, as she entered the English-dominant classroom and was immediately tasked with having to learn a new language and culture while, at the same time, grasping the content of instruction.

In response to these new understandings, the teacher-researcher team strived to create and continually refine innovative pedagogical practices to support the student's learning and to provide her with the contextual conditions needed to gain meaningful access to the social and academic practices of the classroom. This chapter, then, illuminates some of the social, material, and organizational resources that were identified as necessary to achieve more equitable educational experiences for students whose dominant language is one other than English. However, before we turn to our study, we provide a brief overview of the literature concerning issues of equity in relation to linguistically diverse students.

Equity and the Linguistically Diverse

Dating from the time of the Civil War, the United States as a society has held a long historical tradition of impelling standardization as a tool for national unification and of driving forth myopic ideologies as the de facto norm in spite of the presence of various social, racial, linguistic, and cultural demarcations. This tradition has often been confounded with the exaltation of nationalistic sentiments, particularly during historical periods of rising immigration rates and, thus, of accentuated xenophobia. For example, during World War I, pervasive Americanization efforts succeeded in banning the use of the German language in many parts of the United States and eventually led to a dramatic decrease in foreign-language learning countrywide (Schmid, 2001). Over time, the push toward standardization has created a hierarchical matrix involving languages, races, ethnicities, social class, and power, which often favor White, middle-/upper-class, monolingual, and monodialectical populations as those who "possess" English. Pierre Bourdieu (1991), whose work has influenced some of the conceptions to be uncovered in this chapter, in his explanations about language as the site of social organization and as a form of symbolic patrimony, asserted the following: "The policy of linguistic unification favors those who already possess the [dominant] language as part of their linguistic competence, while those who know only a local dialect would become part of a political and linguistic unit in which their traditional competence was subordinate and devalued" (p. 46).

Indeed, in relation to schools, many recent studies have suggested that the preservation of English as the hegemonic language relegates linguistically diverse students (in particular non–Anglo-Saxon immigrants and African Americans) to subordinate roles in the classroom (DaSilva Iddings, 2005; Shannon, 1995) and to positions of inferiority in society (Gutierrez, Baquedano-Lopez, & Asato, 2000; Manyak, 2002; Moll, 2002, Toohey, 1998).

Although many researchers in the area of second-language acquisition have pointed to bilingual models of education as being the most effective in promoting educational success for ELLs (Ovando, Collier, & Combs, 2003; Rolstad, Mahoney, & Glass, 2005; Thomas & Collier, 2002), English-only movements continue to gain strength. To date, twenty-four states in the United States have adopted some form of English-only legislation, and others are presently considering adoption. As a result, most ELLs are placed in instructional programs where only one language is used for instruction: English. Clearly, such movements and policies continue to favor the White middle-/upper-class student and to maintain power structures that sustain and preserve dominant societal groups.

Even at a superficial glance (i.e., based only on the most recent test scores published by the National Assessment for Educational Progress [NAEP]), the academic achievement of the linguistically diverse in U.S. schools reflects the repercussions of a long-lived history of educational inequity for these students. For example, 71 percent of ELLs who were assessed in eighth grade, achieved below basic levels, 24 percent scored at basic levels, and 4 percent scored at proficient levels in reading on standardized assessments. Achievement rates for standardized mathematics assessments are almost identical to the reading assessments' scores.

In this chapter we question what it would take to provide a more equitable educational experience for new immigrant ELLs in English-dominant classrooms. In so doing, we examine the empirical case of one recent immigrant Mexican student (less than one year in the United States), her teacher, and the context of their activities.

This Study

Oakwood Elementary School was located in an affluent rural area in the Southeast United States. The school building was about fifteen years old and was very well kept and well equipped with up-to-date technology. The building housed approximately 400 students, 23 support-staff members, and about 38 certified teachers, 84 percent of whom held a master's degree or higher and averaged about seventeen years of teaching experience. The faculty was composed mostly of White, middle-class, female teachers. The average class size was about eighteen students. All fourth-grade classrooms were examples of what the school called the "Twenty-first Century Model," meaning they had at least seven computers, a large-screen TV, and electronic chalkboards. In accordance with local and school-held English-only policies, all instruction was presented in English. Additionally, the school had chosen to implement a pull-out English as a Second Language (ESL) program with the goals of supporting students' academic achievement and developing English-language proficiency as quickly as possible. The district superintendent and the school administrators were concerned that many teachers in the school had very little or no experience or professional preparation in working with linguistically diverse students, and the school was receiving a sudden and large influx of ELLs. The administrators were particularly concerned with the fourth-grade classes, as the students would be taking national high-stakes tests (e.g., Terra Nova) at the end of the year, and the students needed to be better prepared.

Thus, the motivation for this current study came from the district superintendent and school principal's request to help increase the academic achievement of the ELLs enrolled in four fourth-grade classes. As part of a larger study, this chapter specifically focuses on the social and academic circumstances of one recent immigrant Mexican student as she entered the English-only school environment and interacted with the classroom materials, the classroom teacher, and the other students in the class.

Participants

At the onset of our study, our focal student, Analisa (a pseudonym) had recently emigrated from Jalisco, Mexico, where she had attended school and completed fourth grade. According to her records, she was reading at grade level in Spanish and was a high-performing student in all academic content areas. She had arrived in the United States after school had already begun (September), and school officials decided to place her again in fourth grade because of her emergent levels of English proficiency (as attested by the language assessment scale [LAS]).

The teacher, Katie (also a pseudonym), was a White, monolingual speaker of English and had no experience in teaching ELLs. This was her first year at Oakwood Elementary, but she had about five years of prior experience teaching elementary grades in a private school in the area. She had no formal training to work with linguistically diverse students. However, she was eager to learn more about how to meet the academic and linguistic needs of the students enrolled in her class and was particularly concerned about how best to reach Analisa, a newcomer to the English language and to American culture.

The other twenty-two students in the class were, to a lesser extent, also participants in this study. Four of those students were native speakers of Spanish at various levels of proficiency in English, and the remaining eighteen were monolingual English speakers.

Procedures

Our research at Oakwood Elementary lasted roughly one academic year. During this time we held weekly conversations with the teachers to gain an understanding of their prior knowledge about the learning processes of ELLs, their beliefs about the ways these students learn, and any autobiographical influences that may have affected their knowledge and beliefs about ELLs. In addition, the teachers identified challenges from their own practice, which informed the topics of discussion and the selection of pertinent readings for use in prospective meetings and also prompted coaching and demonstration lessons for the development of systematic practices to aid in effective instruction for ELLs.

For this study, we paid particular attention to Katie's classroom, as she seemed very concerned about how to provide equitable educational circumstances for Analisa without being able to communicate effectively with her. About three times per week we observed Katie's classroom, and, in particular, we focused our observations on Analisa's participation in regular classroom activities such as text read-alongs, classroom discussions, and independent work. All observations were videotaped, and these videotapes were then discussed during debriefing research meetings and during our conversations with Katie. The discussions around the videos contributed to build knowledge of sociocultural, historical, linguistic, and pedagogical factors that mediated Analisa's learning processes. Also, ongoing analysis of teacher diaries, students' assessment portfolios, classroom discourse, and practices provided evidence of both changes in Katie's thinking about the developmental trajectories of her English-learning students and, more specifically, the impact of pedagogical innovations on Analisa's ability to access the social and academic practices of the classroom—an issue that emerged as one of the greatest challenges the teacher faced.

Findings

The data to be presented here are from field notes taken while observing Analisa performing regular classroom activities, notes taken during our conversations with Katie, videotapes of classroom practice, classroom artifacts (i.e., student journals, samples of student work, and attitude assessments), theoretical notes used to reference our developing research focus, and anecdotal records of researchers' debriefing sessions.

Although we recognize that, traditionally, critical theory (á la Frankfurt school) is a more common lens for the investigation of issues concerning educational equity and social justice, for this research we were primarily interested in the interrelationships among the social, material, and organizational resources that mediated Analisa's linguistic and academic development and that enhanced her possibilities for meaningful participation in the classroom. To capture these interrelationships, we based our analyses and interpretations on sociocultural theoretical frames, and more specifically on cultural historical activity theory put forth by Lev Vygotsky and his colleagues, A. N. Leont'ev and A. Luria.

Activity Theory

Influenced by Marx and Engels's conceptions of dialectical materialism, and recognizing that human cognition is a material, practical process, with objective (not purely abstract) properties, Russian theorists such as Vygotsky and his colleagues (Leont'ev and Luria) argued that humans use symbolic tools, or signs (e.g., numbers and arithmetic systems, music, art, and language), to mediate and to regulate our relationships with others and with ourselves (Lantolf & Thorne, 2006). As such, Vygotsky pointed at the analysis of signs, and especially language (in a broad sense), as the minimal unit in which the form, function, and genesis of consciousness could be investigated. As these signs are cultural and historical in nature, this perspective came to be known as cultural-historical psychology. Leont'ev (1978), adding to Vygotstky's framework, proposed the observation of *action* that occurs on both interpersonal and intrapersonal levels (i.e., tool-mediated action) as a proper unit of analysis, emphasizing that human development emerges in the social plane and is mediated by the *activity* in which we engage (i.e., activity theory).

In the West, this theoretical approach has been extended by the work of scholars such as Cole (1999), Engeström (1987), and Wertsch (1985), who have applied it to understanding organizational change and development through a synthesis of Leont'ev's activity theory and Vygotsky's cultural-historical psychology—an approach known as cultural-historical activity theory (CHAT). A cultural-historical activity theoretical framework describes social mediation within the larger framework of community, rules, and division of labor. It provides a methodological tool for exploring the dynamic interactions between the various activity systems that exist within a given community (e.g., the classroom). Our meaning for "methodological tool" here is in agreement with Newman and Holzman's (1993) articulations of activity theory as method, which is "simultaneously prerequisite *and* product, the tool *and* the result of the study" (Vygotsky, 1978, p. 65) and not with more traditional understandings of method as a tool that, when applied, yields results (Holzman, 2006). In addition, it recognizes the agency of individual learners collaboratively constructing knowledge within practices and moving toward the object (intention) of activity. Importantly, the elements that compose activity are not seen as static or as existing in isolation from each other but are seen as dynamic and continuous in their interaction with each other. Furthermore, within activity there are contradictions or tensions among the elements. These tensions have been characterized by Engeström (1987) as double-bind situations (for example, in the case of an English-monolingual teacher and a Spanish-monolingual student who need to communicate with each other in order to fulfill their roles in the class but do not speak the same language). These double-bind situations prompt participants to "innovate, create, change or invent new instruments for their resolution through experimentation, borrowing or conquering already existing artifacts for new uses" (Engeström 1987, p. 165 as quoted in DaSilva Iddings & McCafferty, 2005). From this perspective, individuals and materials can transform in largely unpredictable ways, including ways that may otherwise be suppressed or marginalized in institutional systems.

For our work here, we have chosen to analyze three classroom events characteristic of everyday practices of the classroom we observed, which helped us understand the transformative processes that were taking place, both moment to moment and over time. These transformative processes involved the social, material, and organizational affordances that enhanced Analisa's possibilities to participate meaningfully in the classroom context and thus served to legitimate her role as a student and ultimately to create more equitable educational circumstances in the classroom.

Analysis and Interpretations

At the onset of our study, Analisa was often seen in the peripheral margins of the classroom, doodling or simply staring at a computer screen, seemingly daydreaming for long periods of time. Katie felt she was continuously trying to engage Analisa in the content of instruction but was not succeeding. She began to think of Analisa as being very passive and unwilling to learn. When asked to reflect on Analisa's participation in class, she commented,"[Analisa] is just very passive. She just isn't very aggressive about her learning. But if I would sit down and work with her, she would do it. And she always does what I ask her to do. But she isn't very eager to learn. If I'd let her sit at her desk all day, that is just what she would do."

Indeed, at various points during the beginning stages of our study, when Analisa was asked by one of the researchers (in Spanish) what she thought the teacher was talking about during instruction, Analisa would shrug her shoulders and continue to tune out whatever was going on around her. However, as the academic year progressed, discussions between the teachers/researchers involving the videos of Analisa's participation in the classroom served as an important mediator for Katie to change her practices and to create pedagogical innovations that ultimately helped Analisa begin to access the content of instruction. These changes and innovations involved interactions between Katie and Analisa as well as between Analisa and other students and/or symbols and materials in the classroom, where those involved in the activity were able to distance themselves from their immediate context and to create a broader context for their activities. This distancing liberated them from the constraints of the particular setting in which they were functioning and enabled them to construct a new activity system (Centre for Studies in Advanced Learning Technology, 2002). In the sections that follow we provide examples of how these systemic reorganizations and developments took place.

Material Resources

At the beginning of our study Analisa was often assigned class work that was different from that assigned to her English-dominant classmates. In all subjects other than mathematics, Analisa was provided material that focused mainly on the basic elements of the English language. Specifically, these materials presented simple vocabulary and alphabetic exercises intended for students of a much lower grade level. While Analisa was completing this work, the other students were engaging in social studies and English-language arts lessons appropriate for the fourth grade. Within a couple of weeks of Analisa's arrival to the class, Katie spoke of her focus on drilling the student with basic principles of English: "I needed to do things that would help her just to learn the language right now. Things that are not so focused on her being able to sit down and do things with the book, or sit down and do things with paper and pencil. So, I started doing phonics things with her and helping her make flash cards and just having her read and talk to me."

However, in the midst of these kinds of daily activities involving simple language drills, Analisa presented her teacher with a request. One day, while her peers were working on a review of material that the class would see on a future standardized test, Analisa asked Katie if she could do the same worksheet as the other students in the class. Katie remembers, "What she wanted to do was to try the [testing] stuff. She came over to me and pointed to it and I asked her did she want to try it; she said, 'Yes,' and I said, 'Ok, tell me you want to try it.' And she said, 'I want to try.'"

From that moment on, Katie began to assign Analisa the same work as the rest of the class, making methodological accommodations but keeping the content of instruction within the same rigor as for the other fourth graders. These accommodations involved an increased use of gestures, pictures, videos, and other kinds of communicative resources that did not depend exclusively on linguistic cues. Importantly, these accommodations also included the use of the Spanish language as a medium of interactions between Analisa and the other native speakers of Spanish in the class, as well as reading materials written in Spanish that were relevant to the various topics of instruction. She also allowed Analisa to write elaborations in Spanish of what she was learning. Katie celebrated, "I've been having her write in Spanish to be able to express herself. And what she's been doing, which I think is really cool, is she'll write one thing for me in English and then she'll write it in Spanish. So, I think that that helps her to make her feel that she's explaining things."

Interestingly, earlier in the year, Katie had been strongly opposed to the use of Spanish in the classroom as she thought the use of a language other than English in the classroom to be a violation of English-only regulations. However, being able to use her native language served as an important resource for Analisa who, in this way, was able to participate in more meaningful ways in the regular practices of her classroom. As a result, almost immediately, Katie noticed a change in Analisa's classroom attitude. Whereas previously Analisa would often sit in a sort of isolation with little to no sign of engagement, she was now beginning to express herself and was much more animated in class. Katie recounted, "You know, now she has a smile on her face—she's happy. I think she's excited that she's doing the same thing as everyone else. And when she does get something right, she's really excited."

In this example, we place emphasis on two particularly important shifts in the activity systems of the classroom. First, Analisa, who had been perceived as passive and unwilling to learn, reclaimed her role as a student and exercised agency by voicing her need for more equitable treatment in the class (i.e., asking to do the same worksheets as the other students in the class). Second, coming to realize that local English-only mandates did not prescribe the language students can use, Katie began to allow Analisa to utilize her native language in order to make sense of the content of instruction. We argue that these shifts in activity were mediated by the introduction of instructional materials that provided Analisa with access to the content of instruction and also permitted more meaningful participation in classroom practices.

Organizational Resources

With regards to organization and structure of curricular content and delivery, in agreement with standards-based reforms, Katie reported following the State standards rigorously and keeping with the pace of the other fourth-grade classes. That is, all fourth-grade teachers planned their lessons together, used the same textbooks, and went over the same units of study at similar times of the day on the same day of the week. While this delivery structure seemed effective in meeting the demands of covering curricular standards, facilitating lesson planning for the teachers, and helping the actual implementation of curriculum, it also established a lock-step approach where all fourth-grade students were expected to master the content at the same time and pace. Within this structure, students for whom English was not a dominant language often did not have much of a chance in keeping up with their counterparts. Such was the case for Analisa.

With that understanding in mind, Katie decided not to follow the curriculum pacing of the other fourth-grade teachers any longer—particularly in relation to science. Aiming to undercut the usual limitations of textbooks and to provide opportunities for more authentic learning experiences that catered to the specific needs of her students, Katie chose to create her own units for science. However, Katie described being frustrated on various occasions when Analisa seemed very confused about the topic of instruction, even if Analisa sometimes was able to correctly complete the worksheets that were part of the unit being studied: "It's just challenging all the way around. I don't know if I'm doing the right thing; I don't know if what I am doing is enough; I don't know . . . Analisa . . . speaks hardly any English, and she is just most challenging because I'm not really sure how much she is learning or what she understands and what she doesn't."

During one of our teacher/researcher sessions where we reviewed the videos of Analisa's participation in a science lesson regarding the thematic unit of study on nocturnal animals, Katie noticed that Analisa had correctly colored the owl, the bat, and the ringtail possum on the worksheet. Subsequently, however, during an interview it became clear that Analisa had not understood what these animals had in common, and she had not grasped the relationships between the animals and the current unit of study.

In watching Analisa struggle with making connections between the tasks she had been asked to do (e.g., coloring certain animals on a worksheet) and the actual purpose of instruction (e.g., studying about nocturnal animals), Katie recognized that, while she had the best interest of her students in mind when she made the choice not to use a textbook, perhaps she had not been fully aware of the particular needs of Analisa. That is, while there are many educational constraints related to the use of textbooks, they often have a clear and explicit structure that may serve as important organizational resources for ELLs at the emergent levels of proficiency (e.g., index, table of contents, headings, subheadings, and so on). After this recognition, Katie became more aware of the need to reorganize the curriculum in order to make the contents of her units of study more accessible to Analisa. She explained, "We did a unit on plants, a plant unit. What I did was, I had another student work with her, and we drew pictures of the plant parts and then we'd say the words to her and have her repeat them and point to the pictures. Since I couldn't give her the textbook to read, I did things like that. I would have her match the plant part to the English word. She would say [it] in Spanish and I would have her say it in English and then kind of do matching."

By the second half of the year, borrowing from the organizational structures of textbooks, Katie created her own advanced organizers for the units she planned, which explicitly related the lessons to the overarching purposes of instruction, provided a clear flow from one lesson to the next, and overtly signaled the relationships between the topics included in the unit and the lessons.

In this example we point out that Katie's dilemma of not wanting to keep a lockstep approach or use textbooks for the instruction of science, but at the same time wanting to provide a clear organizational structure that Analisa (and other students in the class) could easily follow, created a double-bind situation within the system of activity. This double bind was resolved through the experimentation of different formats and through the innovations in ways to organize the curriculum.

We point out also that this reorganization of curriculum served to counteract some of the inherent conflicts of standards-based educational reforms, which suggests that *all* students should master the academic content at the same time (and which by its own definition, "standard" denies the idea of diversity). The systemic reorganizations of the curriculum in that class served as a form of mediation for Analisa to engage with and to make sense of the content covered by the sequence of lessons in a thematic unit. Through this mediation, she was also better able to make connections with her previous learning experiences from the time when she was still in Mexico, since she already had the opportunity to study some of the same content while in fourth grade there. In this way, the systemic changes in curriculum organization also changed Analisa's learning activities in the class and promoted greater understandings of the content area (science). Although Katie recognized that it sometimes continued to be difficult to gauge what kinds of understandings Analisa was gleaning from the units of study, she thought that if she continued to keep the curriculum within the same pace and organization as the other fourth-grade classes, Analisa would be virtually unable to access the kinds of information she needed to make sense of instruction, thus making it difficult to attain the goal of a more equitable educational experience.

Social Resources

At the beginning of the school year, very little interaction between students took place in Katie's class. Most of the time spent in the classroom revolved around teacher-fronted activities because Katie had been very concerned about meeting the demands from school officials to "get through" the contents of the curriculum as quickly and efficiently as possible in order to increase standardized test scores. However, realizing that this social structure of the classroom did not provide many opportunities for Analisa to engage in authentic language practices (in English or in Spanish) or opportunities for her to be included in the social practices of the classroom, Katie began to change some of her pedagogical practices to provide Analisa with various points of entry into the academic content and the classroom community, hoping that these changes would promote greater educational equity: "I've been doing as much as I can [for Analisa] . . . giving her opportunities to interact with the rest of her classmates . . . just anything I can do for her to be more integrated in the classroom. She interacts more often with the other students who speak Spanish. But she does have a couple of the other little girls in the classroom [whom] she has kind of befriended and will talk to them. Their conversations are a little bit limited, but they still play together on the playground and do jump rope."

In this way, Analisa was gradually learning English and becoming more integrated as a student in the class. On one occasion during the second semester of the school year, Katie assigned the class a project involving creating banners that were to be placed all along the walls in order to cheer on the students while they worked on their standardized tests. The students were assigned the task of coming up with their own confidence-building slogans. These slogans included, "You're doing great!" and "Keep it up!" The students were grouped into teams of four or five and placed around the room to work on their banners. Analisa's team included four other students, one of which, Marta, spoke both English and Spanish. The group decided that each member would contribute one slogan to the banner.

When it came to Analisa's turn to offer a slogan, she preferred to pass as she felt unable to devise a clever saying. However, her own perceived inability to create a slogan did not preclude her participation in the group as the other students made a concerted effort to keep her involved in the project. For example, Marta translated the group discussions for Analisa and worked on her slogan with Analisa's input in Spanish. In addition, Analisa took the other students' slogans and decorated the banner. For instance, one of the slogans was "You're a star!" Analisa read the slogan and added a number of stars around the writing itself. During this activity, she continued to work with her Spanish-speaking teammate as well as with her English-speaking peers and to decorate the banner as the slogans were added. When completed, this banner, along with all of the others, was placed on the wall and served as a needed confidence boost to all of the students during test time.

As the academic year progressed, Analisa was becoming much more of a legitimate member in the classroom community and was quickly learning to communicate in English (although she was still considered to be at emergent stages of proficiency). By her second semester in school, she was more willing to raise her hand and to volunteer her answers when Katie asked the class questions. In addition, she was actively attending to the instruction and was much more agreeable to participating in assigned group work, even if she felt that she might not be able to fully contribute to the assignment.

We point out that Katie's role as a teacher (who was under great pressure to improve test scores) was at times at odds with her desire to provide a more equitable educational experience for Analisa. As a result of these fundamental tensions, new pedagogical practices and structures were implemented, which in turn provoked changes in Analisa's ways of being and acting in the classroom. These changes not only created opportunities for Analisa to gain access to the community of her peers but also afforded possibilities for her to engage in language-learning practices as she participated in problem-solving tasks with other students in the class. Also of note is the fact that Analisa's scores on standardized tests in reading comprehension and English language arts increased three grade levels during that one school year. We submit that the developments illustrated by this example were mediated by the changes in the social structures of the classroom, which proved to be an important source of mediation for the attainment of a more equitable and successful educational experience for the student.

Implications and Conclusions

In this section we return to our original question of what it would take to promote equitable educational circumstances for recent-immigrant students in English-dominant classrooms. We begin to address this question by clearly stating that we position ourselves in the field of education as advocates for bilingual education. In addition, agreeing with Bourdieu's (1991) propositions about language as a form of symbolic capital, we believe that true equity cannot exist in classroom environments where only one language, one knowledge base, or one set of values takes supremacy over others. However, with our illustrations above, we hope to have articulated some of the possibilities that may help enable teachers of ELLs in the English-dominant classroom to approximate that goal. That is, if new immigrant children are to be included in English-dominant classrooms, the data presented here suggest that there are supports that are *essential* in order to minimize institutional constraints that may prevent these students from accessing the content of instruction, to become legitimate members of their classroom communities, to develop and expand their linguistic resources, and to discover and extend their identities as students in a U.S. classroom (see also DaSilva Iddings & Rose, 2007).

Some of these essential supports that have been illuminated in this chapter include the following:

- The unrestricted, purposeful use of native language
- Instructional materials that do not compromise the integrity and rigor of the curriculum
- Clear and explicit organization of content
- Abundant opportunities for peer interaction

These recommendations, in some sense, have been consistently present in the literature addressing the academic needs of ELLs. However, in the majority of U.S. classrooms, these supports have not been implemented to any degree. Instead, the current educational accountability reforms have, at their core, a one-size-fits-all education, packaged curricular materials, and high-stakes testing. Herein, the idea of *equity* is superseded by the prevailing call for conformity (i.e., standardization), and the dynamism of educational contexts is suppressed by the static rules that are forced upon them.

For example, in relation to our study, we note that the context of the classroom, the teacher, and the students were constantly affecting and being affected by the collective transformations that were dynamically taking place moment to moment and over time in that educational environment. However, mandates and policies that were external to that activity system, such as the ones that encompass English-only regulations, had been imposed onto the classroom context without taking into consideration the actual context of activity. In this way, these regulations were fundamentally static, functioning largely as a barrier rather than a guiding principle. We point out, also, that Katie's goal of providing a more equitable educational environment only became more possible once she began to push the boundaries of some of these regulations thus creating, along with Analisa and the other students in the class, new rules, tools, and organizations, which ultimately led to the creation of an ever-renewed context of activity.

Traditionally, direct causality for the high indices of school failure for ELLs has been assigned to the inadequacy of individual administrators, teachers, parents, or the students themselves. Conversely, in this chapter, our motivation in analyzing the educational circumstances of one recent immigrant student through an activity-theoretical lens has been to make evident the complex interdependence of people and their context of activity and to focus on the dynamic configurations and the multifaceted patterns of classroom practices. These are circumstances that, if understood more as a system-like array of relationships, do not have to be inherently deterministic but can, to some degree, be generative of possibilities and transformations. Thus, beyond the curricula and pedagogical implications already mentioned in this chapter, this kind of analysis also has implications for professional-development programs for teachers of English-language learners. The awareness of the intricate interrelationships between specialized teaching and the learning processes of language learners requires the acknowledgement of the agency of individuals. Researchers must also continuously search for specific means to support these students' rights to learn and to achieve academic success.

References

Bourdieu, P. (1991). *Language and symbolic power.* Cambridge: Cambridge University Press.

Cole, M. (1999). Cultural psychology: Some general principles and a concrete example. In Y. Engeström, R. Miettinen, and R. Punamaki (Eds.), *Perspectives on activity theory* (pp. 19–38). Cambridge: Cambridge University Press.

DaSilva Iddings, A. C. (2005). Linguistic access and participation: Second language learners in an English-dominant community of practice. *Bilingual Research Journal, 37*(3), 223–236.

DaSilva Iddings, A. C., & McCafferty, S. G. (2005). Language play and language learning: Creating zones of proximal development in a third-grade classroom. In A. Tyler (Ed.), *Georgetown University round tables on language and linguistics: Language in use* (pp. 112–122). Washington, DC: Georgetown University Press.

DaSilva Iddings, A. C., & Rose, B. C. (2007, April). *The impact of an in-situ, research-based teacher development program on the achievement of reading comprehension for Spanish-speaking English language learners.* Paper presented at the annual meeting of the American Educational Research Association, Chicago, IL.

Engeström, Y. (1987). *Learning by expanding an activity theoretical approach to developmental research.* Helsinki, Finland: Orienta-Konsultit.

Centre for Studies in Advanced Learning Technology (CSALT; Producer). (2002). A video interview with Yrjö Engeström. Retrieved on September 14, 2006 from CSALT, Lancaster University, UK, Web site: http://csalt.lancs.ac.uk/alt/Engestrom

Gutierrez, K., Baquedano-Lopez, P., & Asato, J. (2000). English for the children: The new literacy of the old world order, language policy, and educational reform. *Bilingual Research Journal, 24*(1), 87–116.

Holzman, L. (2006). What kind of theory is Activity Theory? *Theory and Psychology, 16*, 5–11.

Lantolf, J. P., & Thorne, S. L. (2006). *Sociocultural theory and the genesis of second language development.* New York: Oxford University Press.

Leont'ev, A. N. (1978). *Activity, consciousness, and personality.* Englewood Cliffs, NJ: Prentice-Hall.

Manyak, P. (2002). "Welcome to salon 110": The consequences of hybrid literacy practices in a primary-grade English immersion class. *Bilingual Research Journal, 27*(2), 421–442.

Moll, L. (2002). The concept of educational sovereignty. *Penn GSE Perspectives on Urban Education, 1*(2), 1–11.

National Clearinghouse for English Language Acquisition (NCELA). U.S. Department of Education. (2006). *ELP Standards and Assessment.* Retrieved on June 10, 2007 from http://www.ed.gov/about/offices/list/oela/index.html.

Newman, F., and Holzman, L. (1993). *Lev Vygotsky: Revolutionary scientist.* New York: Routledge.

Ovando, C. J., Collier, V. P., & Combs, M. C. (2003). *Bilingual & ESL classrooms: teaching in multicultural contexts* New York: McGraw Hill.

Rolstad, K., Mahoney, K., & Glass, G. V. (2005). The big picture: A meta-analysis of program effectiveness research on English language learners. *Educational Policy, 19*(4), 572–594.

Schmid, C. L. (2001). *The politics of language.* New York: Oxford University Press.

Shannon, S. (1995). The hegemony of English: A case study of one bilingual classroom as a site of resistance. *Linguistics and Education, 7*, 175–200.

Snow, C. E. (2002). *Reading for understanding: Toward a research and development program in reading comprehension.* Santa Monica, CA: RAND Reading Study Group.

Thomas, W. P., & Collier, V. P. (2002). *A national study of school effectiveness for language minority students' long-term achievement.* Santa Cruz, CA: Center for Research on Education, Diversity, and Excellence (CREDE).

Tienda, M., & Mitchell, F. (2006). *Multiple origins, uncertain destinies.* Washington, DC: National Academies Press.

Toohey, K. (1998). Breaking them up, taking them away: ESL students in grade 1. *TESOL Quarterly* 32(1), 61–84.

Trueba, H. T. (1999). *Latinos unidos: From cultural diversity to the politics of solidarity.* New York: Rowman & Littlefield.

Vygotsky, L. S. (1978). *Mind in society.* Cambridge, MA: Harvard University Press.

Wertsch, J. (1985). *Vygotsky and the social formation of mind.* Cambridge, MA: Harvard University Press.

4

As Cultures Collide

Unpacking the Sociopolitical Context Surrounding English-Language Learners

Jason G. Irizarry, University of Connecticut, and
John W. Raible, University of Nebraska–Lincoln

The voices of English-language learners (ELLs) and their families and communities are often marginalized in the discourse regarding the education of linguistic-minority students. Moreover, efforts aimed at improving educational outcomes for ELLs are often rooted in a deficit perspective that views linguistically diverse students of color as inferior and undesirable. Therefore, in an effort to foreground their voices and, at the same time, to critique inaccurate assumptions regarding this community, our chapter utilizes an unorthodox methodology to ensure the inclusion of frequently silenced voices. Our approach draws from cultural sources including print media, documentary films, personal experience, family stories, and academic research. Our approach parallels the "counter-story" methodology described by Solórzano and Yosso (2002) as a "tool for exposing, analyzing, and challenging majoritarian stories" (p. 23) by centering the experiences and voices of marginalized peoples. In this chapter, we forward the subaltern voices of current and former ELL students, as well as their family and community members, with the goal of contributing to a more complex understanding of ELLs and the sociopolitical context of their education. As multicultural teacher educators, we concur with Les Back (1996) that "these muted voices must be integrated into any understanding of the contemporary politics of culture and identity" (p. 6) and, further, that today's teachers must understand their work in relation to ELLs' personal and collective struggles, for example, for cultural survival, educational equity, and social justice.

We realize that we are wading into tricky waters. While we are not bilingual educators or English as a Second Language (ESL) teachers per se, we approach this work as former classroom teachers and current teacher educators and researchers concerned with improving the quality of education for culturally and linguistically diverse communities. As we see it, given the highly communicative nature of teaching and the importance of linguistic communication in the relationship between teaching and learning, every teacher is a language teacher. Building on the work of scholars in the field of multicultural education (Delpit & Dowdy, 2002; Gay, 2000; Ladson-Billings, 1994, 1995; Nieto, 2004), we assert that language, literacy, and culture are inextricably linked. Our intention is not to overwhelm readers with a paralyzing portrait of the dire straits confronting ELLs and those who work with them. Rather, we hope to forward, through the use of empowering personal voices, a view of ELLs and the families and communities they come from as vibrant, rich, and vital entities that do, in fact, assert agency and challenge the larger sociopolitical context of their schooling.

Our goal as multicultural teacher educators is to prepare teachers, including preservice teachers and those already in the field, to enter their sites of practice with *eyes wide open*, fully cognizant of the complexities and profound responsibilities of the work they have chosen to do. Rather than attempting to romanticize the facts, we opt for candor with the intent of demystifying a situation that remains, all too often, clouded by ill-conceived deficit models that place the blame on ELLs and other young people instead of on the institutions that seek to control them. We hope readers will come away not only more aware but also more committed to facing the challenges of providing meaningful educational experiences for today's ELLs and other students of their generation.

History Informs the Current Context for Learning and Teaching

> Your grandmothers and grandfathers spoke in Spanish until the day they died. Everybody we knew did. Sure, they could speak a little English now and then at the store or whenever they needed to, to get by. But they were proud *Hispanos*. They traced their lineage back to España, back to Spain. They could show you the documents signed in the royal court that gave their families their tracts of land in New Mexico way back in the 1500s. As far as they were concerned, Spanish was the national language. People who speak today of making English the national language don't know what they are talking about. Our *familia* has always lived on Spanish-speaking soil. It is our land—*nuestra tierra*—and our way of life. We've been on this land longer than the Pilgrims, longer than those Anglo so-called founding fathers. And we speak Spanish and a little English. Don't you ever forget it, *m'hijito*. Don't let anyone tell you different. Be proud of who you are. Know where you come from, know your history. Most of all, love your *familia*, love the land, and love *la Raza*.
>
> —Voice of an Elder (Voice #1)

As multicultural educators, we draw from the groundbreaking work of Sonia Nieto (2004) and the more recent scholarship of Nieto and Patty Bode (2008) to define what we mean by the "sociopolitical context" of education. As we understand it, the sociopolitical context in which ELLs are situated, and thus defined, is rooted in the history of protracted cultural collisions that flared up in the contest between the colonizers of the Americas. The literal wars between England, France, Spain, and other European powers—not to mention against the indigenous peoples who were overwhelmed by imported diseases and intentional genocide—underwrite the origins of the present-day culture wars. We define the contemporary culture wars following Ira Shor (1988) to describe the systematic rollbacks undertaken largely by political conservatives to dismantle the gains of the left-leaning social movements of the mid-twentieth century. These rollbacks include, among numerous others, attacks on reproductive rights for women and legal protections for sexual minorities (such as gay men, lesbians, and members of the transgender community), state-by-state attempts to outlaw progressive interventions (such as affirmative action and same-sex marriage), as well as more localized efforts to water down (i.e., depoliticize) multicultural education and dismantle bilingual education in its various forms. We hold that today's ELLs are caught in the crossfire of raging debates over immigration reform, the future of bilingual education, and mean-spirited (and short-sighted) English-only initiatives, all of which represent the continuation into the present day of age-old conflicts about social and economic control over and within colonized territory. In our view, the current educational status of ELLs is made all the more precarious due to its outgrowth from the nexus of centuries-long collisions between cultures and distinct ways of life.

Current political debates around issues such as bilingual education and immigration tend to overlook or simply ignore the legacy of physical colonization as well as linguistic imperialism that have been at the center of U.S. expansion. From our perspective, Spanish speakers, who currently constitute the largest linguistic-minority population in the United States, have become an increasingly popular target in the culture wars. Spanish speakers, otherwise known as *hispanohablantes*, number approximately thirty-two million, representing almost 12 percent of the entire population (U.S. Census Bureau, 2005). After English and Spanish, the next most common language spoken in the United States is Chinese (all variations), with two million speakers. Therefore, one can argue that efforts to suppress linguistic diversity are reflective of "hispanophobia" and represent a form of linguistic colonialism, which, we argue, is central to understanding the current sociopolitical context in which ELLs are embedded.

Much of what is currently referred to as the Southwest of the United States was once northern Mexico. While the Treaty of Guadalupe Hidalgo (1848) resulted in the imposition of a new border between the United States and Mexico, many individuals still refer to this land as "occupied America" (Acuña, 2000) or *Aztlán*, the ancestral home of the indigenous Nahua peoples. Referring to the territory in these terms signals not only to the contentious struggle for land and the vestiges of colonialism but also references how these factors contribute to shaping the identities of many of the lands' current inhabitants.

While individuals of Mexican descent comprise the largest percentage of *hispanohablantes* in the United States, Puerto Ricans, the second largest group, share a similar history of colonization, first by the Spanish and later by the United States. The island of Puerto Rico, a U.S. colonial possession since 1898, is one of the oldest colonies on the face of the earth. When the United States first occupied the island, one of its first imperialistic acts was to change the mode of instruction in schools from Spanish to English (Negrón de Montilla, 1975). Puerto Rican teachers who were not fluent in English were forced to teach using a language that neither they nor their students understood. Language was, and continues to be, a primary vehicle for the cultural suppression and subjugation of Puerto Rico.

The cultural collisions that were a product of U.S. imperialism and westward expansion in the eighteenth and nineteenth centuries played out in schools on the U.S. mainland as well. Mexicans and Mexican Americans in the Southwest/*Aztlán* at the turn of the twentieth century were often denied access to any education. In fact, by 1915 nearly 80 percent of Mexican children in Texas did not attend school, and even as late as 1944 only approximately half of school-aged Mexican children were enrolled in school (Cockcroft, 1995). Opportunities for formal education that were available were typically in segregated schools, housed in dilapidated facilities, with outdated and often biased curriculum materials that were discarded by white schools. Such conditions reflect majority views of and prejudice against "undesirable" populations.

Contrary to misconceptions and stereotypes regarding the educational aspirations of Latino families, Mexican parents in the Southwest/*Aztlán* have a long legacy of activism and struggle for educational equity and social justice. For example, responding to a lack of access to quality education for their children, parents in Texas withheld the names of two hundred school children from U.S. Census personnel, resulting in significantly less money for per pupil expenditures flowing into the district. If their children could not access the same resources as their white counterparts, parents decided they would not subsidize the education of Anglo students. While many people today are familiar with the 1954 landmark Supreme Court decision that ended *de jure* segregation in the United States, few are aware that the first successful school integration case was won by Mexican Americans in Lemon Grove, California, in 1931 (Alvarez, 1986; Garcia, 2001). Upset by the fact that their children could not have access to the new school constructed in the town for white students, Mexican American parents mobilized members of their community and brought a case against the school district, eventually winning the first successful school-integration case in the history of the United States. Despite this early victory, Latinos were not officially recognized as covered under the *Brown v. Board of Education* decision until 1973 (Cockcroft, 1995), and today they remain one of the most segregated group of students in the country.

Beyond access to education and the quest for equality of resources (including facilities, books, and qualified teachers) that were assumed to accompany the promise of racial/ethnic school integration, was the issue of quality classroom instruction. Because many Mexican American families spoke Spanish as their primary language, schools in the Southwest/*Aztlán* were faced with the task of educating students who spoke languages most teachers did not understand. Although there is a current backlash against using any language other than English as a means of instruction in public schools, it is important to note that the Treaty of Guadalupe Hidalgo, which officially ended the U.S. war with Mexico, recognized the cultural distinctiveness as well as the linguistic and educational rights of Mexicans remaining in the Southwest/*Aztlán* after the U.S. acquisition of close to half of the land belonging to Mexico. Moreover, early state constitutions in states such as New Mexico explicitly outlined provisions for bilingual education in these newly acquired states (MacDonald, 2004). According to the Bilingual Education Act of 1968 and the Supreme Court decision in *Lau v. Nichols*, instruction for ELLs has to be meaningful. That is, it is unconstitutional (i.e., illegal) to immerse students in classrooms where they can understand little or none of the linguistic communication in the classroom. That being said, there continues to be a significant discrepancy between legal and educational policy and practice. There are many published stories of linguistic-minority students, past and present, languishing in classrooms where they understand little if any of the language spoken in the classroom.

While this chapter tends to focus on *hispanohablantes* as the largest linguistic-minority group, we remind readers that contemporary ELLs include students from Native communities, both in urban and reservation settings. Native Americans long have felt the effects of colonization and deculturalization. The imposition of English-only schooling has been a key weapon in the government's struggle to "civilize" American Indians, Alaskan Natives, and Hawaiian islanders during the past two hundred years (Spring, 2004), and, to this day, struggles for cultural survival, including language revitalization projects and bilingual education, continue in various Native communities.

Contemporary Collisions and the Attempt to Control "Undesirables"

> So, if you really want to hurt me, talk badly about my language. Ethnic identity is twin skin to linguistic identity—I am my language. Until I can take pride in my language, I cannot take pride in myself . . . Until I am free to write bilingually and to switch codes without always having to translate, while I still have to speak English or Spanish when I would rather speak Spanglish, and as long as I have to accommodate the English speakers rather than having them accommodate me, my tongue will be illegitimate. I will no longer be made to feel ashamed of existing. I will have my voice: Indian, Spanish, white. I will have my serpent's tongue—my woman's voice, my sexual voice, my poet's voice. I will overcome the tradition of silence.
>
> —Voice of a Scholar (Voice #2): Anzaldúa, 1987, p. 59

A wealth of scholarship in the area of language acquisition suggests that it takes students approximately five to seven years to develop meaningful proficiency in another language (Crawford, 1999; Cummins, 2000). Transitional bilingual-education programs, aimed at transitioning students into "mainstream" English-dominant classrooms, were typically structured to provide ELLs support for their academic growth and language acquisition for up to five years using the students' native language in classroom instruction. In 1988, President Ronald Reagan, despite the wealth of evidence to the contrary, mandated that transitional programs be limited to three years. More recently, voters in California, Arizona, and Massachusetts have abandoned transitional models in favor of one-year "sheltered immersion" programs where students are allowed limited support in their first language for one year and are then immersed in English-only classrooms, in essence eliminating promising practices in bilingual education in these states with significant ELL student populations.

As the nation continues to grapple with relatively recent legislation such as school desegregation, as evidenced by remedies implemented in response to the 1954 *Brown* and 1974 *Lau* Supreme Court decisions, ELLs continue to be positioned as political pawns whose educational needs vie for attention against the objectives of more powerful forces.

Our multicultural, subaltern perspective holds that history renders it too simplistic to accept the majority narrative of a "nation of immigrants" expanding under the watchful and welcoming eye of a benign Statue of Liberty. U.S. history involves far more turmoil and intercultural strife than the mainstream patriotic narrative suggests, which prefers to tell its story as a pleasant melding of peoples from diverse cultures, all of whom arrived with open hearts and minds and a generosity of spirit to pursue religious freedom and economic security for their families. As Voice of an Elder #1 reminds us, large parts of what is now U.S. territory literally belonged to other nations (both indigenous and European) for generations, if not centuries. Moreover, given this chapter's focus on ELLs, it is worth noting that the North American lands now controlled by the United States have always been multilingual (Loewen, 1995). Thus, the story of today's ELLs is not simply a tale of newcomers who "don't speak English." It is a deeper and more complex story of ongoing conflict and conquest, of economic and political winners and losers, and of cultural hybridity and moments of intercultural cooperation. A more accurate narrative would include stories of wholesale genocide and transnational movements of people, both in the United States and abroad, spurring concerted efforts for cultural survival, mass migration, and immigration, including to North American shores. In some present-day communities in the United States (for example, in parts of the Southwest/ *Aztlán*), historical conflicts rage on, if only on an emotional or ideological level. Families and communities continue to struggle with issues of ethnic pride and identity in which language issues play a significant role.

Another example of the struggle for cultural affirmation and survival unfolded more recently in Oakland, California, when in 1994 the school board proposed using Ebonics (also referred to as African American English Vernacular) in the teaching of African American students. The argument advanced by the Oakland School Board was rooted in the premises that students' cultural sensibilities, including language, should be affirmed and that the culture of African Americans can potentially be used as a bridge to help students acquire the "codes of power" (Delpit, 1995) necessary to successfully navigate schooling. The Oakland proposal received criticism from those opposing any form of bilingual education as well as from critics rejecting the idea that Ebonics is a distinct language that merits consideration in the classroom. Much of the antipathy toward Ebonics, as well as other language forms, emanates from oppressive historical relationships (between the powerful elite and the "undesirable" populations) that continue to manifest themselves in the present day. The rejection of Oakland's proposal to make instruction more culturally responsive to the cultural and linguistic needs of African American students, along with the electoral victories in several states to outlaw bilingual education clearly demonstrates the intense political nature of English-language learning. To remedy this situation, teachers need to understand their work in classrooms as necessarily a *political* activity and not only as an educational one.

Language Teaching as a Political Act

> Language learning is the process by which a child comes to acquire a specific social identity. What kind of person should we bring into being? Every vested interest in the community is concerned with what is to happen during those years, with how language training is to be organized and evaluated, for the continued survival of any power structure requires the production of certain personality types. The making of an English program becomes, then, not simply an educational venture but a political act.
>
> —Voice of a Scholar (Voice #3): Rouse, 1979, p. 2

We believe that a significant part of our responsibility as multicultural teacher educators is facilitating the development of what Lilia Bartolomé and Donaldo Macedo term "political clarity" (Macedo & Bartolomé, 2000, p. 4). One aspect of that project involves offering our teacher education students a variety of perspectives from the study of history, politics, and literature to counter the popular tendency toward an ahistorical view of intercultural relations in our pluralistic society. From this perspective, all teaching is understood as a political act that takes place within a complex, though often unexamined, sociopolitical context.

To this end, we help teacher education students to understand the work of schools in what Ira Shor (1988) refers to as the current era of "conservative restoration." Shor identifies the beginnings of the restoration with the rightward political turn in the 1980s with the election of Ronald Reagan as president. The conservative restoration was the latest installment of the culture wars between defenders of the gains won by the popular social movements of the mid-twentieth century (for instance, the civil rights movement, women's liberation, and gay liberation) and the mainstream establishment that had lost cultural and political ground. The "wars" were understood to be cultural insofar as they involved pitched debates around questions of culture—that is, definitions of what counts as knowledge, which canons of knowledge would be taught in schools and universities, whose language would dominate as the official language, and even what were the definitions of cultural institutions such as marriage and the family. Specifically targeted by the Right were many of the progressive innovations in education, including ethnic and women's studies, affirmative action, and multicultural and bilingual education.

Shor's cogent analysis of the culture wars explains clearly how the nation's rightward shift in the 1980s marked the beginning of what would become sustained assaults on hard-won progressive gains. According to Macedo and Bartolomé (2000), there remains today a direct correlation between the colonial legacy of white supremacy, current grabs for global corporate power, and calls for anti-immigrant policies on the domestic front. Under the recent U.S. regime, the Bush administration's No Child Left Behind Act can be understood as one of the latest accomplishments of the conservative restoration. Explicating the links between the current discourse of accountability, high-stakes testing, and cultural hegemony, Joel Spring (2004) argues, "The NCLB Act represents a victory for those advocating that schools teach a uniform American culture . . . The primary emphasis in the legislation is on the acquisition of English rather than support of minority languages and cultures" (p. 123).

Whereas progressives advocated for more inclusive schools with increased multicultural curricular offerings, antibias structural reforms, and expanded opportunities for historically marginalized students (including ELLs), the conservative restoration has systematically attempted to curtail the inroads made by the social movements of the 1960s. In their insightful review of Macedo and Bartolomé, R. Truth Goodman and Kenneth Saltman (2001) explain how the restoration relies on an explicitly racialized discourse: "Conservatives have launched a discourse on cultural and linguistic purity, insisting that White culture, White power, White jobs, White language, and White territory are endangered because Black and brown people are claiming pieces of 'our' precious tax base, national sovereignty, and heritage" (p. 27). Clearly, schools remain one of the main battlegrounds upon which the culture wars continue to be waged, as majority discourse decries the dilution of traditional values, "dumbed down" canons, and increasingly "impure" communities.

Returning briefly to 1980s scholarship for additional historical and contextual insight, we draw upon the perspective of Native scholar and author Vine Deloria, Jr. In his prescient article *Identity and Culture* (1987), Deloria took stock of the last hundred years of social and political struggles in various ethnic communities. Concluding his essay by looking to the future of interethnic relations, Deloria suggested that as members of minority communities became homogenized through education and the rapidly expanding influence of the media, once-vital subcultures might lose their vigor and distinctness: "Racial minorities will always have a social/political identity because of their obvious presence within a large white majority . . . Brutal measures have declined with the increase of majority sophistication, but basic attitudes create new forms of oppression, unique to the age but ancient in impact. As the cultural traditions of racial minorities erode and become homogenized by modern communication, the fearful possibility exists that these groups will be sapped of their natural resources for survival and become perpetual *wards of the welfare state*" (in Takaki, 1987, p. 103, emphasis added).

While the above passage, which was first published in 1981, comes across today as rather prophetic, what Deloria could not have foreseen was the rapid rise in the 1990s of the numbers of actual (as opposed to metaphorical) wards of the state. Nor could he likely have predicted, to quote proponents of neoliberalism, the "end of welfare as we know it." For example, Deloria could not have foreseen the staggering growth of what has been described as the prison-industrial complex (see Angela Davis, 1999), which has added exponentially to the ranks of wards of the state in real terms. Vince Beiser, writing in *Mother Jones* in 2001, documented the astronomical rates of incarceration in the United States during the 1980s and 1990s. Beiser contends that "the number of Americans held in local, state, and federal lockups has doubled—and then doubled again. The United States now locks up some two million people . . . And the number is still growing" (Beiser, 2001). The United States ranks number one among industrialized nations in the percentage of the population it incarcerates. Beiser further documents how, since the 1980s, per capita spending on prisons grew six times faster than spending for higher education. Moreover, the racial disparity of prison populations has been widely publicized. That is, the clear overrepresentation of inmates of color has become common knowledge. We concur with Beiser's analysis that the reasons behind these alarming rates are not only escalating rates of crime. On the contrary, as Beiser asserts, "It's not crime that has increased; it's punishment. More people are now arrested for minor offenses, more arrestees are prosecuted, and more of those convicted are given lengthy sentences. Huge numbers of current prisoners are locked up for drug offenses and other transgressions that would not have met with such harsh punishment 20 years ago" (p. 63).

Along with the rapid rise of the prison industrial complex, other social phenomena provide examples of the ongoing culture wars and, in our view, further attempt to position so-called undesirables as wards of the state. For example, we can look to recent attempts to assert official government agency in the effort to control language use as evident in recent legal and civil court cases. In several instances, court judges, acting in the role of representatives of the state, have used their platforms to rebuke parents for speaking languages other than English to their children, going so far as to threaten dire consequences for noncompliance. For example, in a 1995 Texas divorce and parental-custody case, a state district judge ruled that it was "child abuse" for a Mexican-born mother to speak Spanish exclusively to her child, suggesting she would be relegating her child to a life of servitude as a "housemaid" (Verhovek, 1995). In another example, a Tennessee judge instructed a mother of Mexican descent who was in court in part for failure to immunize her child (arguably a controversial practice), to "use birth control and learn English" (Barry, 2005). In yet another case, the same judge threatened a Mexican mother with terminating her parental rights if she did not learn to speak English at a fourth grade level within six months, that is, in time for her next court date. According to the *Los Angeles Times* article about this case, the court order noted that "the court specifically informs the mother that if she does not make the effort to learn English, she is running the risk of losing any connection—legally, morally and physically—with her daughter forever" (Barry, 2005, p. A14). The threatened removal of children from their unfit and undesirable (i.e., non–English-speaking, non-Christian, and "uncivilized") parents harkens back eerily to the widespread roundup of Native children and the origins of the Indian boarding-school movement of the late-nineteenth century (see Spring, 2004).

The ward-of-the-state metaphor can be applied further to describe the experiences of a Mexican woman who was held in a Kansas hospital against her will for a decade, spanning from 1983 to 1993. The hospital staff diagnosed and treated the woman as a schizophrenic after attempting to converse with her in English and Spanish. However, the woman's primary language was Raramuri, an indigenous language similar to that spoken by the Aztecs. A lawsuit filed on the woman's behalf stated, "All of the defendants [which included more than a dozen doctors and several other professional staff] forced Ms. Quintero to abandon her ethnic identity and conform to Euro-American cultural customs by forcing her to attend certain activity therapies, by forcing her to change her behavior, her dress, her language, her thoughts, her beliefs" (Corrin, 1996).

Efforts to usurp the rights of linguistic minorities and to exert increased control over their communities are clearly evident in the actions of the aforementioned government agents. In the first example from Texas, speaking to a child in a language other than English was equated with child abuse. In the Tennessee case, not only did the judge attempt to infringe on a woman's reproductive rights by suggesting that a mother appearing in his court not bear any more children, but the court also threatened to physically remove another woman's child from her custody if she could not meet its prescribed level of English proficiency prior to the next scheduled hearing. The final example tells the story of the forced incarceration and concurrent culturally insensitive treatment of a woman based solely on a psychiatric diagnosis made as a result of conversations in a language she did not understand. Although each of these events is shocking in and of itself, it is important to consider these cases not as random acts but rather as present-day examples of institutional oppression against ELLs with historical roots that date back centuries.

While not necessarily related to efforts to increase the ranks of legal wards of the state, we find further evidence to increase state control over certain groups of deemed "undesirable" in the movement to make English the official language of the United States. Beginning in the early 1980s, groups such as U.S. English have advocated for the creation of a constitutional amendment to identify English as the country's official language. Cloaked in benign efforts to "unify" the country around one language, the potential outcomes associated with the creation of such a national policy are far more sinister. One of the most significant implications of this proposed legislation would be a mandate to ensure that all federal documents, including election ballots, would be available only in English. Given the proficiency needed to navigate the process of voting, it is likely that many ELLs would become disenfranchised as a result. Considering the recent significant population gains made among minority populations in the United States, coupled with future population projections that predict the deepening ethnic, racial, and linguistic texture of the country, perhaps these actions are not unintended consequences.

Efforts to mandate one official language fail to recognize the multilingual history of the country. Written in 1777, the Articles of Confederation, the country's first constitution, were disseminated in English, French and German (Crawford, 1999). However, more than two centuries later, as the United States has become increasing linguistically diverse, proponents of English-only legislation believe government documents should be available in only one language. In fact, there is evidence to suggest the authors of the Constitution deliberately refused to name an official language, suggesting that choice, and in this case choice related to language, should be the cornerstone of democracy (Brice-Heath, 1992; Crawford, 1999). Certainly proposed English-only legislation presents serious implications for all ELLs; given the predominance of people in the United States who speak Spanish as a primary language (the United

States is home to the second largest Spanish-speaking population in the world), it would have a disproportionate, adverse affect on the burgeoning population of *hispanohablantes.*

Anti-immigrant initiatives and proposed language restrictions, as Macedo and Bartolomé (2000) point out, have become a key trope in a vote-getting discourse based largely on racism and fear of the Other (or what we are calling "undesirables") in the popular imagination. Once again, we see how ELLs, many of whom come from immigrant communities, are positioned as pawns in ongoing sociopolitical controversies, rather than as individual children and youth with pressing educational needs.

From Collision and Containment to Cultural Connectedness

> My family came here from Cambodia when I was real little. I'm 14 now; I was a baby then. My parents work very hard so us kids can get an education and have a better life here. A lot of my friends are in a gang. It's hard because they are my friends, you know, but I see that thug life is not for me. I'm trying to keep myself out of trouble, but trouble always seems to find me. I been suspended three times this year, mostly for attendance, once for fighting. I was defending myself after we got jumped by these American boys. It makes no sense. They say go to school, but if you don't, we gonna throw you out anyway. What kind of logic is that? If I don't show up in school, I'm in trouble, but if I do show up in school, I'm in trouble, too. My teachers don't like me. They do not understand me or respect me or what I'm about. They think because they can't speak my language or even pronounce my name that I must be dumb or less than them. I am afraid that if I get suspended one more time, I may get deported. Yes, actually sent to Cambodia because I was born there. My probation officer told me about this. They got new rules about that. You mess up here, you find your ass over there in some place where you don't know anybody or how anything works. It happened to one of my homies. I don't want to lose my family and everything I know just because of one slip I happen to make.
>
> —Voice of Adolescent ELL (Voice #4)

The voice of the Cambodian ELL teenager above expresses the legal limbo facing a number of ELLs from immigrant families and the high stakes facing others who find themselves occasionally in legal trouble. Henry Giroux (1997) has alerted educators to the adversarial legal predicament facing not only young ELLs, but the entire generation now experiencing childhood, adolescence, and young adulthood. According to Giroux, adult society has not only turned its back on traditional "undesirable" populations but also has expanded the parameters delineating undesirable populations to include *all* youths: "Releasing itself from its obligations to youth, the American public continuously enacts punishment-driven policies to regulate and contain youth within a variety of social spheres" (p. 2). In Giroux's view, adult society equates youths as a class with the blurring of traditional lines of difference, and it writes off many youths as undesirable: "The attack on youth and the resurgence of a vitriolic racism in the United States is, in part, fueled by adult anxiety and fear over the emergence of a cultural landscape in which *cultural mobility, hybridity, racial mixing, and indeterminacy* increasingly characterize a generation of young people who appear to have lost faith in the old modernist narratives of cultural homogeneity, the work ethic, repressive sublimation, and the ethos of rugged individualism" (Giroux, 1997, p. 6, emphasis added).

Blaming youths for many social ills impacts ELLs to an even greater degree, as evidenced in the example of the teen whose voice opened this section (see Voice #4). Even as the aging generation of adults is apparently fascinated by the more spectacular aspects of youth culture (e.g., youthful celebrities and athletes), it nevertheless positions youths as a problem to be solved through more and more regulation and containment. Poor youths and youths of color, in particular, bear the brunt of this ageist assault. California's Proposition 21 (the Juvenile Crime Initiative of 2001) epitomizes the ways in which adult society has come to fear and then criminalize the young, just as the national discourse around immigration reform demonizes and seeks to punish undocumented workers as "illegals." The California example is especially pertinent to our discussion of ELLs since it occurred in a multilingual state with broad pockets of burgeoning multicultural and "minority-majority" populations, particularly in its major cities.

A study of Californian urban youths' identities conducted by Fazila Bhimji (2004) provides insight into the divide-and-conquer dynamic in this adult-driven legislation. The Juvenile Crime Initiative effectively separates unruly youths into two camps, namely "good kids who make mistakes" and "bad kids" who are dangerous and incorrigible and therefore in need of incarceration (p. 40). Bhimji describes the harsh nature of Proposition 21, which not only allows children as young as fourteen to be tried in adult court for murder and serious sex offenses but also increases the powers of prosecutors to try young people as adults for less serious crimes, among other provisions. In Bhimji's analysis, "The representations of urban youth, reflected in California's policies and in the media, portray urban youth of color as deviant, dangerous, and morally un-reformable" (p. 43).

In both instances, that is, in the case of California's Proposition 21 and in the immigration-reform debate taking place at the national level, official calls for *containment* provide politically expedient responses to the artificially manipulated sense of moral panic and adult insecurity. Containment can happen on many levels: through building a wall along the border between two nations, through self-appointed vigilantes who patrol the border, and through the construction of more and more prison cells. The discourse of containment also attempts to control unruly youths discursively through policing their language, music, and even dress styles (think of the growing popularity of uniforms in urban *public* schools, as well as adult indignation over the use of profanity and the widespread use of the "N-word" in hip-hop lyrics). Linking the criminalization of youths as a class to fear of what we are calling "undesirables," Giroux (1997) writes, "The growing demonization of youth and the spreading racism in this country indicate how fragile democratic life can become when the most compassionate spheres of public life—public education, health care, social services—increasingly are attacked and abandoned" (p. 15).

For youths in immigrant communities, including many ELLs, the stakes now include deportation to countries with which they may have no cultural or linguistic familiarity, particularly if they were born in the United States or came to the United States as very young children. As teacher educators, we feel obligated to prepare preservice teachers for an admittedly harsh sociopolitical reality that is most effectively addressed without sugar coating. At the same time, we feel responsible for engendering a spirit of hopefulness so that novice teachers will stay in the trenches and take up arduous and challenging, but nonetheless rewarding, work with young people. In the final section, we turn to possibilities and practices that we believe can make a difference for today's ELLs, particularly when teachers view themselves as allies in struggle.

From Pedagogy to Empowerment

> This is not a pretty picture of the world being constructed for us by the market system. These are toxic conditions for our work in colleges. What to do? . . . Connect to others who are in the same boat. Full and part-time teachers, along with our students and their families, need to stand together to get the . . . programs we deserve, but we also have to stand with the cheap labor that cleans our classrooms, dorms, and offices, as well as with the cheap labor that grows our delicious bananas in Guatemala and stitches our elegant running shoes in Vietnam. The choices are only becoming more stark—solidarity or inequity, solidarity or barbarism.
>
> —Voice of a Scholar (Voice #5): Leo Shor in Parascondola, 2007

In this chapter, we have argued that ELLs, along with other populations that are deemed threatening and undesirable in majority discourse, are positioned in marginalized and controlling ways. We have shown, through examples of the actual voices of ELLs, their family members, and scholars writing from within and about ELL communities, various ways in which ELLs have responded to marginalization and pathological positioning. In concluding this chapter, we are mindful once again of how the teaching of ELLs should be the responsibility of all teachers, not only those designated as ESL or bilingual educators. We draw this conclusion in the realization that because teaching and learning are fundamentally communicative, linguistic, and therefore cultural enterprises, *each* of us has an obligation to attend to issues of language and culture in education.

We have further argued that one important aspect of preparing future teachers is encouraging teacher-education students to develop political clarity about the larger sociopolitical context surrounding their chosen profession. We have drawn on the definition of sociopolitical context in which teaching and learning occur as advanced in the work of Nieto and Bode (Nieto, 2004; Nieto & Bode, 2008), particularly where ELLs are concerned. In some ways, given today's political climate during the conservative restoration (Shor, 1988), all teachers who work in public schools can be said to have been relegated to the ranks of the undesirables by being marginalized themselves. If we equate the majority view of the teaching profession with marginalized communities, particularly in public schools that brim with increasingly diverse students, we can then draw inspiration and strength from the creative responses to marginalization on the part of other supposedly undesirable groups, including ELLs.

In our view, teachers must eventually understand their work as Paulo Freire (1970) understood the work of teaching, particularly when working with ELLs and others from cultural- and linguistic-minority communities. That is, we hope to encourage preservice teachers to see their role as standing not over but also *alongside* their students in ongoing solidarity to challenge the marginalization thrust upon all of us. In this sense, teaching can be reinterpreted as a political act of solidarity and in service to the ideals of social justice and cultural change.

We draw hope from our experiences in these communities and draw inspiration to continue to champion this cause within our work as multicultural teacher educators. Our optimism is renewed again and again by the courageous examples of various members of ELL communities who, in the face of great odds against them, continue to struggle for educational equity, social justice, and cultural survival.

One step teachers can take is to develop strong alliances with the communities from which their ELL students come. Gil Conchas (2001) also underscores the need for teachers to develop strong ties to students and their communities when teaching students from cultural- and linguistic-minority communities. In his report on his study of variability among urban Latino students, and how one particular school constructed school failure and success, Conchas emphasized the need for schools to "structure learning environments that link academic rigor with strong collaborative relationships among students and teachers" via the implementation of supportive institutional and cultural processes (p. 502).

Another step teachers can take is to become familiar with the discourse of globalization. As one of the three pillars of globalization (the first, according to Suárez-Orozco, 2001, being new information and communication technologies and the second being the emergence of global markets and postnational, knowledge-intensive economies), immigration and displacement require further study by educational researchers. After all, as Suárez-Orozco writes, "Globalization is the reason that immigrant children are entering U.S. schools in unprecedented numbers. Furthermore, their life chances and future opportunities will be shaped by globalization" (p. 345). An understanding of globalization's impact on teaching and learning must accompany any serious consideration of the sociopolitical context of education for ELLs.

Helping preservice teachers to appreciate and value the diverse communities in which their students live and learn requires teacher educators to assist them in redefining not only their personal identities but also their conceptualizations of home, particularly through the lens of transnationalism. Loukia Sarroub (2001) reminds us that home is not only the space we occupy but also "a set of relationships and ideas that proffer a different set of expectations than those of school" (p. 391). For the Yemeni American girls in Sarroub's ethnographic study, home "constituted a set of relationships among people both in the [United States] and in Yemen." At the same time, never feeling truly at home in either Yemen or the United States, these particular students/sojourners "found a 'home' in managing their liminal space" (p. 413). Sarroub draws on the classic sociological construct of the "sojourner" to indicate a bicultural, transnational individual who "remains attached to his or her own ethnic group while simultaneously living in isolation and staying apart from the host society" (p. 392). Such are the complexities of identity and home under globalization and conditions of postmodernity. Living as they do in two worlds, sojourners are marked by their journeys back and forth between host country and homeland. Teacher educators can cultivate an understanding of the benefits of similar cultural border crossing by providing opportunities to engage in cultural border crossing when working with preservice teachers. Particularly for students who come from dominant or mainstream groups, understanding the complexities of liminality and hybrid identities may invite education students to learn to feel at home in more than just their own culture.

We call for teacher-education students to become, as Nieto discusses, not only multicultural educators but also multicultural people. Just as

ELLs grapple with border crossings on a daily basis, their teachers can meet them halfway. As Suárez-Orozco (2001) points out in his essay on globalization, immigration, and education, "For many today, the issue is managing the complexities of belonging both 'here' and 'there': The 'average' youth enrolled in the Los Angeles Unified School District crosses multiple epistemic, linguistic, and political spaces every day. He is likely to have breakfast in Spanish with his parents, listen to hip-hop music with his African American classmates in the bus to school, and hear about the New Deal in Standard English from his 'White' social studies teacher" (p. 361, footnote 19).

In their case study of U.S.-born adolescents at school, Raible and Nieto (2003) found the same dynamics at play. Many students today—not just ELLs—grapple with finding a sense of belonging, which can be particularly challenging when they embody increasingly complex, and multiple, identities. Teacher educators can provide mechanisms by which preservice teachers can be invited to become more "culturally connected" to the children and youths with whom they will one day work (Irizarry, 2007). One way to foster cultural connectedness is to encourage preservice teachers to familiarize themselves thoroughly with the communities and families of their students. In the process, preservice teachers may become not only aware of but also actually *value* what López (2001) refers to as the "subjugated" forms of parental involvement that previously went unrecognized (p. 434). However, in the process of working to create strong ties of solidarity and caring, teacher educators should be careful not to replicate the pitying "*ay bendito*" syndrome, referencing the Spanish-language phraseology.

Breaking the Silence: Asian American Students Speak Out, a film made by Roberta Wallitt (2004), documents a recent conversation between Tibetan and Cambodian youth from refugee families living in upstate New York. The combined comments of this group of young people, several of whom dropped out of high school, reflect the poignancy of the sociopolitical context in which many ELLs find themselves. One young woman, whose family emigrated from Cambodia, had this to say about her educational situation: "I think education is important, even though I am a drop-out. I might be a loser in other people's eyes, but I am happy." This young person has gone on with her schooling, receiving a general education degree and is now pursuing higher education at a local community college. She added, "Teachers have to have their mind open, and realize that they are learning, too. Parents and teachers have to realize that there's an outside world that students have to deal with. They have to give the students respect and realize that they have harder stuff to deal with in the outside world" (Wallitt, 2004).

We close this chapter with the words of another youth in this group of immigrant students, a young man whose family emigrated from Tibet. In Wallitt's film, he described how vulnerable he felt as a newcomer and how he was ridiculed and even described as "retarded" for being an ELL student. The film concludes, as we now do, with a compelling suggestion. All educators would do well to take his message to heart.

> We listen to them all year
> and we learn.
> Some, we don't listen,
> some we don't really care.
> But we do learn from them and respect what they teach us.
> But to show that they really do listen to me,
> I think I would like to give them a quiz
> at the end of the year and see
> what you have learned about me.
> See if you know me
> at all.

References

Acuña, R. (2000). *Occupied America: A history of Chicanos* (4th ed.). Upper Saddle River, NJ: Pearson.

Anzaldúa, G. (1987). *Borderlands/la frontera: The new mestiza.* San Francisco: Spinsters/Aunt Lute.

Alvarez, R. R. (1986). The Lemon Grove incident: The nation's first successful desegregation court case. *The Journal of San Diego History, 32*(2).

Back, L. (1996). *New ethnicities and urban culture: Racisms and multiculture in young lives.* New York: St. Martin's Press.

Barry, E. (2005, February 16). Tennessee judge tells immigrant mothers: Learn English or else. *Los Angeles Times*, p. A14.

Beiser, V. (2001). How we got to two million. *MotherJones.com Special Report.* Retrieved April 12, 2009, from http:www.motherjones.com/prisons/print_overview.html

Bhimji, F. (2004). "I want you to see us as a person not as a gang member or a thug": Young people define their identities in the public sphere. *Identity: An International Journal of Theory and Research*, *4*(1), 39–57.

Brice-Heath, S. (1992). Why no official tongue? In Crawford, J. (Ed.), *Language loyalties: A source book on the official English controversy* (pp. 20–31). Chicago: University of Chicago Press.

Cockcroft, J. (1995). *Latinos in the struggle for equal education.* New York: Franklin Watts.

Conchas, G. (2001). Structuring failure and success: Understanding the variability in Latino school engagement. *Harvard Educational Review, 71*(3), 475–504.

Corrin, D. A. (1996, June 11). Not all Mexicans speak Spanish. *The Kansas City Star*, p. B1.

Crawford, J. (1999). *Bilingual education: History, politics, theory and practice.* Los Angeles, CA: Bilingual Education Services.

Cummins, J. (2000). Language, power, and pedagogy: Bilingual children in the crossfire. Clevedon, England: Multilingual Matters.

Davis, A. (1999). *Prison industrial complex* [Compact Disc Recording]. San Francisco: Alternative Tentacle.

Deloria, Jr., V. (1987). Identity and culture. In R. Takaki (Ed.), *From different shores: Perspectives on race and ethnicity in America* (pp. 94–103). New York: Oxford University Press.

Delpit, L. (1995). *Other people's children: Cultural conflict in the classroom.* New York: Free Press.

Delpit, L., & Dowdy, J. K. (2002). *The skin that we speak. Thoughts on language and culture in the classroom.* New York: The New Press.

Freire, P. (1970). *Pedagogy of the oppressed.* New York: Seabury.

Garcia, E. E. (2001). *Hispanic education in the United States: Raíces y alas.* Lanham, MD: Rowman & Littlefield.

Gay, G. (2000). *Culturally responsive teaching: Theory, research and practice.* New York: Teachers' College Press.

Giroux, H. (1997). *Channel surfing: Race talk and the destruction of today's youth.* New York: St. Martin's Press.

Goodman, T. R., & Saltman, K (2001). Dancing with bigotry. *Educational Researcher, 30*(6), 27–30.

Irizarry, J. G. (2007). Ethnic and urban intersections in the classroom: Latino students, hybrid identities, and culturally responsive pedagogy. *Multicultural Perspectives, 9*(3), 1–7.

Ladson-Billings, G. (1994). *Dreamkeepers: Successful teachers of African-American students.* San Francisco, CA: Jossey-Bass.

Ladson-Billings, G. (1995). Toward a theory of culturally relevant pedagogy. *American Educational Research Journal, 32*(3), 465–491.

Loewen, J. (1995). *Lies my teacher told me: Everything your American history textbook got wrong.* New York: Simon & Schuster.

López, G. (2001). The value of hard work: Lessons on parent involvement from an (im)migrant household. *Harvard Educational Review, 71*(3), 416–437.

Macedo, D., & Bartolomé, L. (2000). *Dancing with bigotry: Beyond the politics of tolerance.* New York: St. Martin's Press.

MacDonald, V. (Ed.). (2004). *Latino education in the United States: A narrated history from 1513–2000.* New York: Palgrave Macmillan.

Negrón de Montilla, A. 1975. *Americanization in Puerto Rico and the public school system, 1900–1930.* Río Piedras, PR: Editorial Universitaria.

Nieto, S. (2004). *Affirming diversity: The sociopolitical context of multicultural education.* Boston: Pearson Education.

Nieto, S., & Bode, P. (2008). *Affirming diversity: The sociopolitical context of multicultural education* (5th ed.). Boston: Pearson Education.

Parascondola, L. (2007). Cheap labor in a world of precious words: What do writing classes produce? (Interview with Ira Shor). Retrieved on July 25, 2009, from http://louisville.edu/journal/workplace/issue7/parascondola.html

Raible, J., & Nieto, S. (2003). Beyond categories: The complex identities of adolescents. In M. Sadowski (Ed.), *Adolescents at school: Perspectives on youth, identity, and education* (pp. 145–161). Cambridge, MA: Harvard Education Press.

Rouse, J. (1979). The politics of composition. *College English*, *41*(1), 1–12.

Sarroub, L. (2001). The sojourner experience of American Yemeni high school students: An ethnographic portrait. *Harvard Educational Review*, *71*(3), 390–415.

Shor, I. (1988). *Culture wars: School and society in the conservative restoration, 1969–1984*. New York: Routledge.

Solórzano, D., and Yosso, T. (2002). Critical race methodology: Counter-storytelling as an analytical framework for education research. *Qualitative Inquiry*, 8(1), 23–44.

Spring, J. (2004). *Deculturalization and the struggle for equality: A brief history of the education of dominated cultures in the United States*. New York: McGraw-Hill.

Suárez-Orozco, M. (2001). Globalization, immigration, and education: The research agenda. *Harvard Educational Review*, *71*(3), 345–365.

Takaki, R. (Ed.). (1987). *From different shores: Perspectives on race and ethnicity in America*. New York: Oxford University Press.

U.S. Census Bureau. (2005). *Percent of people 5 years and over who speak Spanish at home: 2005*. Washington, DC: U.S. Government Printing Office.

Verhovek, S. (1995, August 30). Mother scolded by judge for speaking in Spanish. *New York Times*, p. 12.

Wallitt, R. (2004). *Breaking the silence: Asian American students speak out* [Video cassette]. United States: Equity Matters. Teaching For Change, P.O. Box 73038, Washington, DC 20056.

5

Schooling and the University Plans of Immigrant Black Students from an Urban Neighborhood

Carl E. James, York University

It is generally accepted that social class together with factors such as race, ethnicity, and generation status (i.e., immigrant, first, second, or third generation) of students affect their schooling experiences and educational outcomes. In fact, the social-class backgrounds of students, in terms of family income and parental education, play a significant role in determining the neighborhood in which they reside, the school they attend, and educational resources to which they have access (Brantlinger, 2003; Finn, 1999; Frenette, 2007; Lareau, 2002; López, 2002; Taylor & Dorsey-Gaines, 1998; Weis & Fine 2005). While middle-class students are more likely to do well in the middle-class school system of North American societies; in many cases, working-class students, especially those of immigrant backgrounds, struggle to do the same. However, there are some cases of students with immigrant parents who, despite the social and cultural differences that exist between them and the school system, capably negotiate the school system to attain their educational goals and those of their parents (Anisef, Axelrod, Baichman, James, & Turrittin, 2000; Boyd, 2002; Fuligni, 1998; James & Haig-Brown, 2001; Louie, 2001; Portes & Macleod, 1999; Rong, & Brown, 2001). In this chapter, I focus on the schooling situation and experiences of working-class students of immigrant parents, noting the complex and diverse ways in which social class combines with other factors to affect their educational aspirations and attainments.

I reference the schooling experiences and educational aspirations and achievements of two African Canadian university students, Conrad and Kendra (all names are pseudonyms), who lived in and attended Northdale High School, in a "sub/urban" working class neighborhood of Duncan Park in Toronto, Canada. The "troubled" neighborhood, as the media refer to it, is home to about one hundred thousand people of diverse ethnic, racial, and religious backgrounds, among them—Italians, East Indians and Sri Lankans (or South Asians), Vietnamese and Cambodians (or Asians), Somalis, Ghanians, and Black/African Caribbeans—believed to be the largest proportion of minority residents. The community can be classified as a high-density area of high-rise apartment buildings and townhouses that was established in the 1960s (and continues today) as a "reception area" for the increasing number of immigrants and refugees arriving in the city. The community is a mixture of public housing, subsidized rentals, condominiums—owned and rented—and private family dwellings. Ordinarily, the area would have been considered a suburb of Toronto, but given its characteristics, the area is referred to as an "urban area" (or "inner city") and the schools in it are referred to as "urban schools."

The economic, social, and cultural character of Duncan Park is reflective of many low-income neighborhoods in large North American cities (Hidalgo 1997; James & Haig-Brown 2001; López 2002; Lorinc, 2003; Myles & Feng 2004). Specifically, it is a community populated by minority and immigrant families—many of whom live in impoverished conditions; and even though White people are part of the diversity, the common perception is that it is a "minority community" populated mostly by "Black" people. This characterization speaks to the segregated structure of the society or "place stratification" (Myles & Feng, 2004). Myles and Feng (2004) found that these "immigrant enclaves" as opposed to "ethnic communities" are, in effect, transitional neighborhoods, but due to economic resources—specifically family income—some immigrants, especially Blacks, get stuck in high-density apartment buildings for low-income renters and in public housing. According to Myles and Feng, "immigrant enclaves" are the areas where immigrants with limited resources cluster upon arrival in the host community; "ethnic communities" are those where recent, wealthier immigrants segregate by choice, forming more enduring, culturally homogeneous neighborhoods (p. 2). Myles and Feng also indicated that in Canada generally, the average Black family income was only 79 percent of White family income, whereas South Asian and Chinese family incomes were 85 percent and 91 percent respectively of Whites (p. 11). Further, Black families tended to be younger, more likely to be single parents, less educated than South Asian and Chinese immigrants, and reside in more ethnically and racially heterogeneous neighborhoods (p. 10).

Studies have shown that drug use and trading, crime, and violence are often by-products of the social conditions of low-income, high-density neighborhoods (Hidalgo, 1997; López, 2002). Hidalgo (1997) observed that the pervasiveness of crime and violence, and a fear of losing their children to these negative influences, made families put great efforts into protecting their children from these potential "dangers" in the neighborhoods. These "dangers" serve to stigmatize these communities and concomitantly, as "a normal occurrence," criminalize youths, particularly Blacks, who become part of, as López (2002) writes of minority youth in New York, "the burgeoning prison industrial complex" (p. 75). Sustaining this criminalization of Black youths in the Duncan Park neighborhood is racial profiling or stereotyping based not only on race or color but also on place of origin (non-White is read as "non-Canadian") thereby singling them out for greater scrutiny or differential treatment (Ontario Human Rights Commission, 2003, p. 6), particularly by police. Typically, Black males are profiled as troublemakers, law breakers, drug dealers, criminals, and bad boys (James, 1998), and females as loud, irresponsible, sexually available, "baby mothers," and "girls who are attracted to the bad boys" (Ali, 2003; Gaymes, 2006; López, 2002).

The profile of Black youths, particularly constructed in relation to the working-class immigrant neighborhoods in which they live, help to shape their schooling as well as their own and others' perceptions of their academic abilities, skills, potentials, and possibilities (see Bloom, 2007). In fact, even though publicly funded, schools in neighborhoods like Duncan Park lack the same infrastructure as well as the political and economic resources and supports as schools in middle-class neighborhoods that benefit from the informal parent-driven fundraising and monitoring activities that result in greater resources for students (Frenette, 2007). Further, the high student dropout, suspension, and expulsion rates, as well as the relatively high number of "at risk" students (in other words, those with behavioral problems and who are academically failing), contribute to a reputation of working-class neighborhood schools that makes it difficult for them to attract teachers with the commitment and expertise to work with their students. Some schools in these working-class Toronto neighborhoods place their students in uniforms, have hall monitors (or security guards, sometimes in uniforms) and security cameras, and lock school doors while classes are in session. These "surveillance" routines are seen as a means of helping "deficient" and insubordinate youths learn self-discipline. And where mentorship and role-modeling programs are developed to help address the educational, familial, and social needs of students, ironically in some cases, they further serve to keep students under surveillance. Indeed, as Odih (2002) contends, "these practices serve to subject already vulnerable, disenfranchised working-class males [and females] to increased scrutiny and regulation by government agencies" (p. 99).

In a study a colleague and I conducted with a racially diverse group of twelve university students from a working-class neighborhood in Toronto (James & Haig-Brown, 2001), we found that with the encouragement and support of their parents, participants aspired to receive a university education. They held this aspiration despite typical low teacher expectations; differences between themselves and their teachers due in large part to social class; the inability of schools to respond to their educational needs, interests, and aspirations; and the resulting feelings of alienation from school. Particularly significant was the large role that the geographic and ethnic communities played in their narratives pertaining to their experiences and aspirations. Some of the youths said that because of the "reputation" of the school and the area, their parents and teachers encouraged them to attend high school elsewhere (see Gulson, 2006). While a few youths did as these significant others suggested, most chose to remain at the neighborhood school hoping, as in the case of some of our research participants, that their educational performance and achievements would help to contest the media's images of their community and school as "bad." For that reason, participants aspired to become teachers, social workers, lawyers, activists, and politicians—careers that they perceived were critical to addressing the needs and issues within their community and to helping to change media images. They used terms such as "paying back," "giving back," and "returning the dues" to indicate the obligation they felt toward their parents, families, peers, and community members.

Clearly, community plays an important role in the lived experiences and expectations of young people. With reference to Conrad and Kendra, I examine how their lives within their working-class neighborhood affect their schooling experiences and educational aspirations and attainment. But before doing so, in the following section, I discuss the theories that inform this work. Thereafter, I provide a profile of the participants and then discuss what they said about life in the community and school. I go on to discuss how Conrad and Kendra use and negotiate various forms of capital in order to ensure their educational successes; and I conclude by noting how, in exercising agency, they attained their educational and occupational goals.

Social Class, Social and Cultural Capital, and Community Wealth

In her article on the effects of class on working class youths' transition from urban high schools in New York City to university, Bloom (2007) argues that "it is important to explore the enduring importance of class in explaining the educational trajectories and life chances of poor and working class youth" (p. 344). Social class, Ng (1993) contends, "is not a static category or a thing; it is a *process* that indicates how people construct and alter their relations in terms of the productive and reproductive forces of society, using whatever means they have at their disposal" (p, 51). Related to this process is the set of usable resources and power that individuals derive from the social (including economic, political, and educational) structure of society. These resources, or cultural capital, are recognized as helping to structure individuals' opportunities and possibilities and are commonly used to explain the differences in educational achievements among students. Parents of middle-class and upper-class backgrounds are thought to have the necessary resources, values, and knowledge—in short, the cultural capital—to pass on to their children to enable them to succeed in the society. Specifically, Lee and Barro (2001) found that higher levels of education and higher income of parents positively affected students' performance and academic achievements (p. 467). These parents have the resources to enable their children (starting with daycare) to have an early education—a head start to provide reading materials at home, to develop verbal skills (including how to summarize, clarify, and amplify information), to develop interpersonal trust, and to live in neighborhoods with good schools (Aizlewood & Pendakur, 2005; Frenette, 2007; Lareau, 2002).

By contrast, parents of working-class backgrounds having less income, lower education, and a lesser amount of institutional knowledge are perceived to lack the knowledge, skills, and understanding of the economic, social, and educational system, thus putting their children at a disadvantage (see Anisef et al., 2000; Bloom, 2007; Henze, 2005; Maynes, 2001). Lareau (2002) also found that working-class parents tend to be deferential and constrained in dealing with authority figures such as teachers, police, and government agents and are generally more distrusting of them than their middle-class counterparts (p. 749). Furthermore, feeling a sense of powerlessness, working-class parents tend to intervene very little or not at all into the schooling affairs of their children, because, as Lareau (2002) noted, they often feel that their interventions into schools are ineffective. Exposed to their parents' ideas, attitudes, and interactions with teachers and school administrators, children of working-class parents are likely to develop a sense of distrust and powerlessness, while their counterparts of middle-class parents tend to cultivate a sense of entitlement, hence expecting that schools will be responsive to their needs and interests (Lareau, 2002, p. 770).

Researchers suggest that the class culture of educational institutions is premised on a Euro-centric ideology that further disadvantages students from immigrant-, racial-, and ethnic-minority backgrounds, hence, compounding the efforts it takes for them to pursue and sustain their educational aspirations (Henry & Tator, 2006; Yosso, 2005). In such a context there is usually pressure to assimilate. But while assimilation or acculturation—that is, the "blending" into the new or host social and cultural structures and related practices—might provide youths access to the needed cultural capital to get by in the society, there are also the youths' sense of their parents and community supports and expectations that can enable them to function effectively in the larger society thereby limiting their marginalization. In fact, as Zhou (1997) contends, in the context of increasing pressures on immigrant-minority youth to assimilate into the larger society, family and community supports can make a considerable difference in providing resources that work toward success and nourish the aspirations of the young members of the society (see also Rong & Brown, 2001). Portes and MacLeod (1999) refer to the resources, supports, and values that immigrants receive from their communities as social capital. It is the "banding" together for moral support and using the resources that exist among them to challenge the racism, xenophobia, and discrimination (such as lack of recognition or the devaluing of their abilities and skills) they face to ensure their economic, social, and cultural survival.

Therefore, the educational achievements of students are not determined merely by the cultural capital resulting from membership in the middle class. Indeed, in the case of immigrants, and as Yosso argues with reference to "Communities of Color," there are other forms of cultural capital that marginalized students bring to their schooling that could enable them to succeed if only these knowledges were acknowledged, valued, and made use of (see Milner, 2006, p. 81; Milner, 2007). Yosso (2005) refers to the various forms of capital upon which Communities of Color draw as "community cultural wealth," which she defines as "an array of knowledge, skills, abilities and contacts possessed and utilized by Communities of Color to survive and resist macro and micro-forms of oppression" (p. 77). Yosso identifies six forms of community cultural wealth—aspirational, familial, social, linguistic, navigational, and resistant capital. She writes, "These various forms of capital are not mutually exclusive or static, but rather are dynamic processes that build on one another as part of community cultural wealth. For example, . . . aspirational capital is the ability to hold onto hope in the face of structured inequality and often without the means to make such dreams a reality. Yet, aspirations are developed within social and familial contexts, often through linguistic storytelling and advice . . . that offer specific navigational goals and challenge (resist) oppressive conditions" (Yosso, 2005, p. 77).

Essentially, social class—in combination with race, ethnicity, gender, language, immigrant status, religion, area of residence, and other factors—helps to determine the educational performance, aspirations, and achievements of students. Participating in the inequitable education system stands to be difficult for students without the normalized middle class "cultural capital." Nevertheless, their abilities and efforts to effectively read, understand, and negotiate the system regardless of their differences—in terms of area of residence, class, race, ethnicity, gender, and immigrant status—can open up possibilities and opportunities for them. Young people's capacity to work with and across social and cultural differences can enable them to draw on their "multi-cultural capital"—the abilities, skills, and knowledge gained from their interactions with various communities (in terms of neighborhood, racial, ethnic, class, gender, and so forth)—to assist them in constructing aspirations and accessing opportunities (James, 2005a, p. 219).

This Study

Conrad and Kendra (all names are pseudonyms) were two of twenty second-generation Black students of Caribbean origin, between the ages eighteen and twenty-five years, who participated in a qualitative research study I conducted in the Toronto area between 2001 and 2002. Both were scholarship students attending the same university and had supplemented their scholarship with student loans. In the larger study, I sought to find out how the raced experiences of participants informed their construction, understanding, and articulation of their identity as Canadians and concomitantly their perceptions of their educational and career opportunities and possibilities. The focus was on second-generation and generation-and-a-half youth noting how their experiences traversing the social, educational, cultural, economic, and occupational structures of the Toronto society were mediated by the social and cultural references and expectations of their immigrant parents and the "Caribbean community" in Toronto. Significant here is the fact that Conrad and Kendra lived and attended school in the urban, working-class area of Duncan Park where the combination of neighborhood, school, and, in some cases, family characteristics are known, in many cases, to have played a role in limiting the educational and occupational ambitions and outcomes of students (James & Haig-Brown, 2001; McLaren, 1998).

The stories and descriptions that Kendra and Conrad tell of their worlds are not taken "as potentially 'true' pictures of 'reality.'" Rather their stories opened up for analysis the culturally rich information "through which interviewers and interviewees, in concert, generate plausible accounts of the world" (Silverman 2000, p. 823; cf. Bloom, 2007; Snyder, 2005). As such, their accounts are not mere "descriptions" of the contexts in which they exist but are complex interpretations of their situations, which are mediated by particular social, economic, cultural, and political structures

(Bloom, 2007; Delgado-Gaitan, 1994; López, 2002; Waters, 1999). Further, the stories of these students, collected in two interview sessions over a one-year period (and more recently a conversation to update) illustrate the ways in which they exercised agency and took on a level of responsibility in navigating and negotiating their schooling and community structures. In what follows, I examine how the culture of the school—in terms of things such as the support and encouragement they received from teachers, coaches, and guidance counselors—functioned to sustain their educational ambitions and make university education a reality for them.

The Participants

Kendra

Kendra was a twenty-two-year-old third-year university student who was majoring in history and French while also pursuing a degree in teacher education. She lived with both of her Caribbean-born parents and her two younger brothers in a condominium they own in the Duncan Park area—the area to which they moved about two years before she completed high school. As if to qualify that their condominium living does not represent affluence, Kendra added, "We don't have a car and that type of thing." She described her parents as "strict," especially her father, which meant that she "could not stay out late" at nights with her friends—many of whom had parents who were also from the Caribbean. At Northdale, the neighborhood high school she attended for about two years, she enjoyed "good relationships with [her] teachers"—many of whom encouraged her to go to university. Such encouragement, plus the support and expectations of her parents and friends—most of whom had planned to attend or were already in university—helped to make attending university a reality for Kendra.

In reflecting on her parents' expectations of her, she said, "I think like any other family they just wanted me to go to school, do well and get an education so that I could get a good job and I'd be able to support myself and maybe to do better than they were able to do back in St. Vincent." Her mother's involvement in the schools she attended—through attendance at parent-teacher meetings and as a member of the school council of her high school—not only served to keep her parents informed of her behavior and progress in school but also served to ensure that Kendra did well in school in order to get to university. However, as Kendra explained in the second interview, her father's "noninvolvement" in her schooling activities does not mean that he was uninterested. As she put it, "When I talked about my dad being laid-back . . . I want to make it clear that I don't see him as a deadbeat dad or anything like that; it's just that he did what he could."

While high school for Kendra was "just okay," her "best memories" were of her elementary and middle schools where she was "a good student"—not "a teacher's pet." (She also enjoyed her first year of university.) Aside from the short period in eighth grade when she participated on the girls' basketball team and "did a little bit of track," much of her time and energy were spent on her academic work. She noted that her participation in sports came about because her elementary school teachers felt that her height and "long legs" would make her a good athlete.

Interestingly, "up to Grade 12" Kendra disliked history, and it was reflected in the grade she received (C+ in the tenth grade of the high school she attended earlier). She clarified that her earlier disinterest in history was because of the homogeneity of what was taught. As Kendra put it, "I think the only diverse topic that we talked about was maybe Natives and it was probably in passing . . . I figured that slavery happened here [in Canada], but again it wasn't anything we were taught. I figured that there were contributions made by Blacks, and I kind of had an idea . . . Even during Black History Month the things we were taught were mainly Americanized, so I didn't like history at all."

But it was in twelfth grade that Kendra took a Canadian history course, "liked it," and received an A+; "so," as she said, "it wasn't a hard decision" in choosing history as the discipline in which to major in university. She further rationalized her decision by saying that she had no knowledge of other disciplines such as sociology and anthropology: "I wasn't familiar with those fields, or I probably might have chosen one of them. I don't regret choosing history. I didn't know that you could study, like major in Caribbean studies so I just figured history." Economics was an exception; she had taken an economics course and had declared that as her major on her application to university. Kendra indicated that although she attended information sessions about postsecondary institutions during her final year of high school, she did not remember asking any questions, nor did she seek any help from the school's guidance counselor, so much of her decisions about university were her own. "It was," as she confidently articulated, "just that I knew I was going to university . . . So, I just figured it didn't really matter what I did."

Like Conrad (as will be discussed later), Kendra seemed to live by a work ethic of hard work, determination, and the mindset that anything is possible as long as one puts his or her mind to it. For instance, in response to the question: Do you see anything stopping you from achieving your goals whatever they may be? Kendra said, "Only myself, I guess, because you can always make something possible, but you have to believe in yourself to achieve it, or if something is not going right for you, find ways to change it. I do know that racism is out there, and I've experienced subtle things but not anything that will stop me—at least right now—from going through [with] what I want to do." Kendra went on to say that while she might have experienced racism, the subtlety of it renders it difficult for her to "pick out"; hence her claim that she cannot remember any experiences with racism. She also added, "I don't let that [racism] define me or stop me from what I need to do." She questioned the tendency to focus on racism when talking about the experiences of Black people: "What about people's lives outside of that?" she said. "What about just the regular things? I know to some extent that you can't get away from it just from the fact that racism is always there, but I don't know. It's just that [it is] not everybody's experience . . . I just think that there is more to life."

Kendra was clear about her identity as "Canadian," but she also took pride in her Caribbean heritage. As she put it, "I primarily think of myself as Canadian. Like I said, I am very proud of my [Caribbean] heritage, but I was born here . . . So I would think that my primary identity would be Canadian or Black Canadian." Based on her many visits to the Caribbean, particularly Barbados, she figured that life is better in Canada, especially where educational opportunities are concerned. Colonialism, she said, has affected the Black people there as is the case in North America.

Conrad

Like Kendra, Conrad was a twenty-two-year-old third-year university student who was pursuing a degree in business administration with the hope of entering law school upon graduation. Ultimately, he wanted to work in international relations. Born in Jamaica, he came to Canada with his mother and older sister at the age of twelve and settled in a rented apartment in the Duncan Park area about a year and a half after immigrating. He started school at Northdale in tenth grade. Talking about immigrating to Canada, Conrad said that back then "I was excited because you are leaving [Jamaica], you are changing; so that was exciting, but apart from that, I was indifferent I think." He remembered that, in his first year in Canada, he was asked what he wanted to do when he grows up, he said then, "I want to get my law degree and go back home to Jamaica." But while today he has no interest in returning to Jamaica (he did not say why), he still aspired to become a lawyer—a career interest that he thought to be a product of his years of exposure to lawyers through his mother, a legal secretary. To the question, "Is this the career your mother wishes for you?" Conrad answered, "I don't know what she wants me to be. I think she is pretty much of the mentality—do whatever you want to do, sort of thing."

Schooling in Canada held "good memories" for Conrad, especially his time at Northdale, which was described as "quite good." He went on to say, "I don't know how I would have coped in a school that's predominantly

White." Because of his "love" for Northdale, Conrad regularly returned to talk to ninth-grade students about paying attention to their school work so that they may do well and become successful. These talks with students also served to remind Conrad about his reliance on his high-school teachers to help him understand the Canadian education and social systems, as his mother was unable to do so. Conrad explained, "I mean, my mom understood you go to school, do your homework, you do well," which also meant that he could do whatever he wanted to do. He added, "There is the whole thing of understanding the system, but then there is the basic principles that go along once you understand it—of how to get through it. And I think my mom helped a lot with basic principles of the hard work ethic in terms of getting through it; and the teachers . . . helped [me] to navigate [the system] in terms of, this is what you need to do; this is what you need to look at."

Conrad's decision to major in business has to do with the fact that he "did math well" and had a desire to "get rich." He unabashedly related his "capitalist mentality" to being an immigrant—the idea of coming to Canada to take advantage of the available opportunities. Not surprising then, he identified as Canadian and at the same time expressed the fact that he is also Jamaican. More importantly, he said, "I am a product of Canada in that I've taken advantage of . . . the opportunities that Canada has to offer, which is what Canada is all about." Jamaica, then, was for him the place where he "grew up." Because, as he explained, "growing up in Canada for me, it was just applying all the stuff I learned in Jamaica. I don't think I'll be considered a typical Canadian because everything I did was pretty much using the principles that I learnt in Jamaica. So I don't know if I would say that I grew up in Canada at all."

Conrad could not remember any experience with racism; like Kendra, he said, "I don't allow [incidents of racism and discrimination] to override me . . . I am sure there have been comments made, like I know stuff has happened, but I just can't specifically remember." He went on to suggest, "You can allow things to govern your life or you can deal with them then and just kind of move on. I don't allow things to just pass by. If there is open discrimination or something like that, I'll say whatever is necessary to be said and then move on."

Life in the Neighborhood

While both Conrad and Kendra lived and attended school in the Duncan Park neighborhood, they had very little interactions with their peers outside of school. This is how Conrad explained things:

Conrad: For me it was pretty simple. Like I went to school, went to track, went home and did my homework. Like outside school I had no friends in the area, so I don't think I had the good experience of what it was like to grow up in [Duncan Park]. I was pretty isolated; even within there I was pretty much in my own world, so I didn't really, I guess, associate with a lot of people from the area outside of school.

Carl: Why?

Conrad: Because [of a] combination of being too busy with stuff and school, [and a] combination of not wanting to get in trouble because of its reputation, pretty much those two things.

Carl: What about the reputation of [Northdale]?

Conrad: The reputation? I don't know, I can only think of my experiences. I mean you have kids that got into trouble, probably because of the socio-economic conditions, there is probably more than your typical school but I loved my experience at [Northdale].

Kendra also had very few friends in the area. In fact, she maintained her friendships with peers from her old neighborhood, and it is to them she referred when she talked about having friends already attending and planning to attend university. Kendra's activities outside of school were mainly with her church and island association where she played leadership roles helping to organize youth activities. That Kendra chose not to participate in activities within the neighborhood likely had to do, as in the case of Conrad, with its "reputation" and not wanting to get "into trouble." Consequently, both Kendra and Conrad felt a need to protect themselves from the possible trouble that lurked in the community. Kendra parents', notably her father's, "strictness," as discussed earlier, was their way of protecting her from trouble in the community and ensuring her safety as a female (see also Hidalgo, 1997; López, 2002).

So in their bid to protect themselves, Kendra and Conrad chose not to be "part of" the Duncan Park community where they lived. Conrad admitted to living a "pretty isolated" life, and Kendra looked elsewhere for fulfillment of her social and cultural needs and interests. But they were not "outsiders"—uninformed about the needs and issues in the community and not having an interest in addressing them. In a way, they recognized that they were "part of" the community because they lived and attended school there and, by extension, were implicated in the reputation of the community and school. For this reason, they felt that by trying to be model citizens and engaging in volunteer activities in school—Kendra helped to plan Black History Month and Kwanzaa school events and tutored ninth grade students during her final year of high school; Conrad ran for president of the student council but lost—they were doing their part to change the perception of the community as "bad." According to Conrad, "I think it's important that people see people from Duncan Park who don't fit the stereotypical mode . . . So if you can break that, or if I can help break that perception just by myself, then I have done something." This idea of playing their part to change the reputation of the community is reflective of findings from an earlier study in which participants also indicated that the careers—teacher, lawyer, or community worker—to which they aspired was because they wanted to "give back" to their community that nurtured and supported them to be the university students that they had become (James & Haig-Brown, 2001).

Life in High School

High school was a very positive experience for Conrad, and somewhat less so for Kendra. Both of them remembered with satisfaction their experiences with their teachers, guidance counselors, and coaches who "pushed" them in their work. As Conrad stated, at Northdale, with "90 percent" of the student population being minority and "probably 70 percent" Black, "the moment you show promise, especially some . . . Black teachers would try to help you." Kendra and Conrad indicate that their experiences were different from other Black students whose teachers and guidance counselors—many of them White—operated on stereotypes, which meant that their teachers had low academic expectations of them and therefore streamed them into non–university-path programs and encouraged them into sports (Douglas, Pitre, & Lewis, 2006; James & Haig-Brown, 2001; James, 2005b). In fact, as Kendra insisted, "I just haven't had that experience with teachers like [them] trying to make me have lower expectations and stuff like that; so I don't know if I've been lucky. Yeah, that's interesting, and that's something I just never thought about."

In the case of Conrad, moving to Northdale was a "welcomed" opportunity, for it was there in the process of enrolling at the school that the guidance counselor advised him to take advanced-level courses because she had noticed his very good ninth-grade performance in the general-level courses he had taken. In reflecting on this pivotal point in his educational life, Conrad said, "I had a very indifferent attitude, and I just didn't really care. And so had it not been for the switch, I could have been taking a whole bunch of general courses and not really going anywhere." Talking generally of his experience at Northdale, Conrad went on to say that it was there that "I was able to get in touch with people who took extra initiative because they saw me as a Black male who had potential."

Unlike Kendra, Conrad took full advantage of the opportunities he got through sports. When he was invited by his gym teacher to try out for the track team, he gladly did so, knowing that participating in track was a "way out" of the community, just like basketball student athletes sometimes rely on getting scholarships to study in the United States (cf. Douglas et al., 2006; Frey, 2004; James, 2003, 2005b). Conrad pointed out that his school was known for its good track athletes, many of whom had won U.S. college and university scholarships. This idea of looking to U.S. athletic scholarship as providing a possible "way out" of the neighborhood and a path toward upward social mobility had to do with the fact, as Conrad explained, that "there are more opportunities for people that are Black there [in the United States], because there's a large number of Black students. And while there is discrimination, I think that there [the United States] is better, in terms of community trying to help people that are Black." This notion that the United States offered better opportunities for Black student athletes was well supported by the coaches and teachers at the school, including his coach, Mr. Basil (pseudonym).

While Conrad did not identify Mr. Basil as his mentor or favorite teacher, Mr. Basil was referred to as one of the few teachers who took extra steps to actually make sure that Conrad did well. But it was Mr. Norman, "who is actually not a Black teacher, a White teacher," whom Conrad remembered the most "because he was the pickiest teacher I ever had." And even though many of the teachers who helped Conrad were from the Caribbean and might have been able to appreciate and understand his situation as an immigrant student from the region, it was Mr. Norman who Conrad credited with giving him the educational skills upon which he had come to rely—that is, "skills to see his mistakes and having the ability to correct them." The supports that Conrad received from teachers, like Mr. Norman, as he suggested, had to do with the fact that "they were just good teachers who actually cared about students." Therefore, for Conrad, what was most important was not the race of the teachers but their love for students, their willingness to "go out of their way to teach . . . and seeing students for what they are, and trying to deal with those students for what they are." Such teachers were not always of the same racial background.

Similarly, Kendra found her White teachers more generous and supportive, and for this reason would go to any of them first if she had a problem, even though she had "a good rapport with Black teachers" and "got along" with many of them. Of the Black teachers she said, "I remember just being in school, and I think it's partly why I wanted to become a teacher as well. I found that some Black teachers seemed mean, and I don't know if it was because they wanted to push the Black students harder to make sure that they achieve . . . Actually, I found that the White ones pushed me more to do, like I guess, more controversial topics than my Black teachers, at least they seemed more interested. It wasn't just because I'm a Black student and you are a Black teacher that I would have this overarching bond."

Part of what Kendra and Conrad were communicating is the expectation that their teachers would not see or treat them like "typical Northdale" students or "typical Black" students because they were not. The typical student, as Conrad suggested,

> is someone who is pretty smart but who just has little drive; who does no homework; goes to class probably gets 70s. Some may decide to go to university when they are done, or some decide to go to college or some decide to get jobs. I think a typical [Northdale] student is someone who is pretty smart but just because of socioeconomic conditions, lack of drive, their status, the way they see themselves in the country, they just don't do well, or language barriers [are] another big thing . . . I think I have some of the same thing that your typical [Northdale] students have; but for some reason, whether it was the family support . . . or teachers actually took the time, or my personal drive, I am not sure what. But I think that I am different from the typical [Northdale] student.

Conrad also asserted that students at his school wished to be thought of and treated differently—as students with university ambitions who with the help and commitment of teachers would be able to get to university and eventually move out of the community. Conrad and Kendra's relationships with teachers were therefore built on the degree to which teachers gave attention to their students' educational agendas, showed them respect, maintained contact with them over time, and were responsive to their needs, interests, and aspirations. These were teachers who cared enough to reach out to them and did not simply see them as "good students," as Kendra said, but showed "special interests" in them.

Furthermore, Conrad and Kendra's comments illustrate that race "sameness" does not obfuscate the class differences that exist between teachers and students. According to Conrad, middle-class Black teachers,

> like to think that they face the same ills as lower-class Blacks because they are Black. But that is not the case. Black students are not going to listen to Black teachers from suburbia . . . just because they are Black; because they . . . don't turn on the TV every day and hear something negative about [Duncan Park]. They don't face living in the community where there are people around them who they know sell drugs, who they know have weapons, who they know are dangerous, and living in that community with the expectation, that's the lifestyle, that's an acceptable life. They don't face being third-generation welfare people . . . And they don't face the low expectation that a lot of kids from [Duncan Park] face, and I don't know if they can relate to

> that. They probably can if they get away from the whole idea; "just because I'm Black, I'm going to help." Or "just because I'm Black, I'm better suited to help." I don't know if that's completely true.

It was not sufficient therefore to be a Black teacher, but a teacher who acknowledged the differences between their middle-class "suburbia" existence and the "dangers" their students faced (cf. Hidalgo, 1997; Odih, 2002). Further, Conrad and Kendra expected that teachers would understand, or make efforts to understand, the issues to which students were exposed—drugs, weapons, welfare, low expectations, and media images—that affected their schooling and educational lives and outcomes.

As indicated earlier, Conrad and Kendra were in their third year of university when they were initially interviewed. In this regard, they were asked to reflect on how well they thought their high school prepared them for university. Both of them entered university with academic scholarships and received assistance and support from their teachers—hence the expectation would be that they were fairly well prepared for university. But it was only Kendra who felt she was well prepared for university. She admitted that she did not find university difficult; and her professors were very supportive and encouraged her to go to graduate school. The only complaint Kendra had about university was the lack of diversity in course content and offerings that provided information about Blacks and African Canadians in particular. By contrast, Conrad, a business student, felt that he was not sufficiently or adequately prepared for university, specifically "in terms of the level of difficulty" of his program. He claimed that the university students from York Region (a suburban area north of Toronto), compared with many of his peers from similar urban areas of Toronto as himself, were "generally on a different level." He explained, "I think that York Region is in a different league in terms of their education system. I know that they pay their teachers more. I think family support for people I know from York Region is generally stronger. I think it's just that the school system in York Region is better than the school system south of York Region" (cf. Frenette, 2007). When asked to say what accounts for the differences, Conrad responded,

> I mean one of the things that a teacher said to me once is that bad schools get bad teachers, and I'm still not sure if that's true. I am sure there are a few teachers at [Northdale] who I don't think should have been teaching because they brought actually nothing to the classroom. But I know there are a few other teachers who definitely should have, because they were amazing, they cared about students. So I don't know that's actually true that bad schools get bad teachers. I think that's probably part of it. I think the next part of it is

> just a matter of [other schools] getting different attention than [Northdale] in terms of the type of teachers that want to go there to the type of teachers that actually go there. And I think [it's] the type of students, the student atmosphere, and you can't underestimate the socioeconomic conditions as well in terms of students.

Discussion

The working-class neighborhood where Kendra and Conrad lived and attended high school combined with their parents' support, encouragement, and aspirations for them helped to inspire their schooling efforts, career aspirations, and educational achievements. Although they did not have access to the cultural capital like middle-class students to assist them in negotiating their schooling and educational needs, interests, and aspirations, they effectively utilized the resources—high aspiration, optimism, determination, and family and teacher supports—available to them. In other words, they made use of the multicultural capital fostered through their immigrant families and community to chart a path toward attaining their educational goals. But as they indicated, the issues they encountered in school and community were not merely related to their identity as children of immigrant parents but also related to race that they understood to intersect with class, immigrant status, and gender. In this regard, they named racism as a contributor to the social and economic situation of people within their community. But interestingly, neither Conrad nor Kendra recalled any personal experiences with racism; while they admitted that it exists, they seemed to think that they managed to circumvent its impact because of their positive thinking, their belief in themselves, and not allowing it to "define," "override," "stop" them, or "govern their lives." It seems, then that Conrad and Kendra coped with racism by telling themselves that if they do not focus on it then they will survive its impact. Their responses to racism could be seen as a form of opposition or resistance to its marginalizing effect—"resistance capital" (Yosso, 2005). This is possibly how they maintained the "hope in the face of structured inequality" that Yosso (2005) suggests is part of aspirational capital.

Yosso (2005) also talks of "navigational capital"—the skills to maneuver through social and educational institutions—that the students from Communities of Color draw on to sustain their high levels of achievement (p. 80). So what may be perceived as Kendra and Conrad creating distance between themselves and the community peers by not socializing with them outside of school is likely a way of navigating the "dangers" of the neighborhood that, if they did not exercise such care, might have operated to disrupt their schooling and educational ambitions. This "distance" that Conrad and Kendra maintained from their peers might have also helped to give their teachers the impression that they were "different" from the other Black students, whom Conrad described as "lacking the drive to do well." It is likely then that their "model minority" status—in Conrad's case, "a Black male who had potential"—appealed to teachers, especially White teachers, who needed to see and work with students who applied themselves to their academic work. Besides, Conrad and Kendra might have successfully used their navigational and multicultural capital to negotiate their schooling context, specifically building relationships with "supportive" teachers, but, paradoxically, it might have also contributed to maintaining the stereotypes of their peers and the community (cf. Landsman & Lewis, 2006; and Milner, 2006 for their discussion of this tendency). Thus, the Euro-centric middle-class structure of the school remained intact and not challenged because Conrad and Kendra were able to "fit in" and succeed through individual efforts. It would be interesting to explore how their constructed "difference" affected the schooling lives of their Black peers. Their successes, such as it is, are likely what have them returning to the school, invited by their former teachers, to talk to students. Such invitations are likely premised on the idea that they are "good role models"—especially Conrad as a male—and individuals who are able to effectively negotiate social, economic, and cultural structures and barriers to attain "meritorious" success on the basis of their own abilities and skills (James & Taylor, 2008)

This notion of "model students" or "good role models" can unwittingly operate as a problem for students who are perceived as "lacking the drive to do well." For as long as students like Conrad and Kendra exist in urban schools, then teachers "with good intentions," as Milner (2006) would say, will probably come to see the problems of low academic achievement, alienation from school, disruptive behaviors, and dropout as more related to the inability of students (and their parents) to meet the expectations of teachers and to handle the school program. What is significant here for teachers is that while students like Kendra and Conrad are prepared to distance themselves from their community peers in order to satisfy the expectations of their parents and teachers, other students might not be willing to do the same because they perceive the cost of doing so—that is, losing friends and disconnecting from the community—as being too high.

Based on her study of Caribbean youth in New York City, López (2002) found that there are no "essential" differences between males and females; nevertheless, there were "differing race-gender outlooks [that] arise due to differences in experiences, perceptions and responses to racialization and gender(ing) processes, not biology" (p. 69). In the Toronto context, as in New York City, as a Black male, Conrad had to contend with the general stereotypes of Black males in the city as those who are athletically—rather than academically—oriented, low achievers, troublemakers, and behavioral-problem students (James, 1998). And as a resident of Duncan Park, there are the additional stereotypes of Black males as gang members, drug dealers, and gun users. In resisting or working against this racializing, gendered profile and employing his resistance capital, Conrad worked hard to demonstrate that these stereotypes did not apply to him, and, in this regard, he abandoned the idea of earning an athletic scholarship—something that he was invited to think about by his coach. Further, with his constructed "social difference," he was able to have teachers see his academic potential. Unlike the males in López's (2002) study who were somewhat ambivalent about their educational prospects, Conrad, like Kendra, remained strongly optimistic about education as a means of upward social mobility and a way out of the community (see also James & Taylor, 2008; Noguera, 2003; Warren, 2005).

The experiences and perceptions of Conrad and Kendra remind us that to deliver an education that is relevant to the students of working-class communities and schools like Duncan Park, attention must be paid to the economic, social, and cultural context in which they live and through which they understand their possibilities in life (see also Brantlinger, 2003; Milner, 2006; 2007; Taylor & Dorsey-Gaines, 1998). As such, the challenge for teachers, particularly middle-class educators and administrators irrespective of their race and other related experiences, is to understand how their identities, class privileges, and expectations of their students influence the schooling situation (Landsman & Lewis, 2006). As Conrad and Kendra have indicated, good, caring teachers—that is, teachers who "go above and beyond the call of duty"—are ones who inspired them and made schooling relevant to their needs and interests (see Milner, 2007). Such teachers do not necessarily share the same race, minority status, or experiences but understand the complex ways in which the selves they bring to the teaching/learning process are informed by their personal histories and perceptions of the students and communities. "Caring" educators also provide space for students to articulate their needs, interests, and aspirations based on their respective social situation and their complex relationship with the community (Finn, 1999; Milner, 2006; Preskill & Jacobvitz, 2001). The classroom, therefore, cannot be seen as separate from the community; however complicated, distanced, shifting, and uneasy the relationships students might have with their community, it remains the context that informs their engagement with school and, as Kendra and Conrad have demonstrated, plays a role in their educational and occupational plans. For instance, although Kendra and Conrad did not feel fully part of the community because they never shared the "mentality" of many of their peer group members, they, did feel a sense of duty to "give back" to their community and help others who are in the same situation they were once in—it is "a natural impetus," according to Conrad.

Conclusion

If schools are indeed to be responsive to the needs, interests, expectations, and aspirations of their students, then attention must be given to the complex and diverse ways in which students engage schools and teachers and exercise agency. In exercising agency, Conrad and Kendra engaged in constructing a personal path toward attaining their educational and career goals in relation to the sociocultural, economic, and community forces that were operating around them. In the absence of substantial parental knowledge of the schooling system, they relied on their own resources and abilities and on their reading of the social, cultural, economic, educational, occupational, and community structures to create a path to university. They maintained that their own efforts, initiatives, and actions, in concert with those of their parents, teachers, coaches, guidance counselors, educational programs, and relevant curriculum, accounted for the social and educational paths they took, the aspirations they constructed, and their educational achievements to date. Today, Kendra is a high-school history teacher, having graduated with a master's degree in history, and Conrad is articling in corporate law, having completed his LLB and MBA degrees.

References

Aizlewood, A., & Pendakur, R. (2005). Ethnicity and social capital in Canada. *Canadian Ethnic Studies, 37*(2), 77–102.

Ali, S. (2003). "To be a Girl": Culture and class in schools. *Gender and Education, 15*(3), 269–283.

Anisef, P., Axelrod, P., Baichman, E., James, C., & Turrittin, C. (2000). *Opportunity and uncertainty: Life course experiences of the class of '73.* Toronto: University of Toronto Press.

Bloom, J. L. (2007). (Mis)reading social class in the journey towards college: Youth development in urban America. *Teachers College Record, 109*(2), 343–368.

Boyd, M. (2002). Educational attainments of immigrant offspring: Success or segmented assimilation? *The International Migration Review, 36*(4), 1037–1060.

Brantlinger, E. (2003). Who wins and who loses?: Social class and student identities. In M. Sadowski (Ed.), *Adolescents at school: Perspectives on youth, identity, and education* (pp. 107–126). Cambridge, MA: Harvard Education Press.

Delgado-Gaitan, C. (1994). *Consejos*: The power of cultural narratives. *Anthropology & Education Quarterly, 25*(3), 298–316.

Douglas, B. B., Pitre, E. D., & Lewis, C. W. (2006). African American student-athletes and White teachers' classroom interactions. In J. Landsman & C. W. Lewis (Eds.), *White teachers/Diverse classrooms: A guide to building inclusive schools, promoting high expectations, and eliminating racism* (pp. 177–149). Sterling, VA: Stylus.

Finn, P. J. (1999). *Literacy with an attitude: Educating working class children in their own self interest.* Albany, NY: State University of New York Press.

Frenette, M. (2007). *Why are youth from lower-income families less likely to attend university? Evidence from academic abilities, parental influences, and financial constraints.* Ottawa: Business and Labour Market Analysis, Statistics Canada.

Frey, D. (2004). *The last shot: City streets, basketball dreams.* New York: Houghton Mifflin Company.

Fuligni, A. J. (1998.) Adolescents from immigrant families. In V. C. McLoyd & L. Steinberg (Eds.), *Studying minority adolescents: Conceptual, methodological, and theoretical Issues* (pp. 127–144). Mahwah, NJ: Lawrence Erlbaum.

Gaymes, A. M. (2006). Making spaces that matter: Black females in public education. Master's thesis, York University, Toronto, Ontario, Canada.

Gulson, K. (2006). A White veneer: Education policy, space and "race" in the inner city. *Discourse: Studies in the Cultural Politics of Education, 27*(2), 259–274.

Henry, F., & Tator, C. (2006). *The colour of democracy: Racism in Canadian society.* Toronto: Harcourt Brace & Company.

Henze, R. C. (2005) Veronica's story: Reflections on the limitations of "support systems." In L. Pease-Alvarez & S. Schecter (Eds.), *Learning, teaching and community* (pp. 257–276). Mahwah, NJ: Lawrence Erlbaum.

Hidalgo, N. M. (1997). A layering of family and friends: Four Puerto Rican families' meaning of community. *Education and Urban Society, 30*(1), 20–40.

James, C. E. (1998). "Up to no good": Black on the streets and encountering police. In V. Satzewich (Ed.), *Racism and social inequality in Canada: Concepts, controversies and strategies of resistance* (pp. 157–176). Toronto: Thompson Education Publishing.

James, C. E. (2003). Schooling, basketball and U.S. scholarship aspirations of Canadian student athletes. *Race, Ethnicity and Education, 6*(2), 123–144.

James, C. E. (2005a). Constructing aspirations: The significance of community in the schooling lives of children of immigrants. In L. Pease-Alvarez & S. Schecter (Eds.), *Learning, teaching and community* (pp. 217–233). Mahwah, NJ: Lawrence Erlbaum.

James, C. E. (2005b). *Race in play: Understanding the socio-cultural worlds of student athletes.* Toronto: Canadian Scholars' Press.

James, C. E., & Taylor, L. (2008). "Education will get you to the station": Marginalized students' experiences and perceptions of merit in accessing university. *Canadian Journal of Education, 31*(3), 567–580.

James, C. E., & Haig-Brown, C. (2001). "Returning the Dues": Community and the personal in a university-school partnership. *Urban Education, 36*(2), 226–255.

Landsman, J., & Lewis, C. W. (Eds.). (2006). *White teachers/Diverse classrooms: A guide to building inclusive schools, promoting high expectations, and eliminating racism.* Sterling, VA: Stylus.

Lareau, A. (2002). Invisible inequality: Social class and childrearing in Black families and White families. *American Sociological Review, 67*(5), 747–776.

Lee, J-W., & Barro, R. J. (2001). Schooling quality in a cross-section of countries. *Economica, 68*, 456–488.

López, N. (2002). Race-gender experiences and schooling: Second-generation Dominican, West Indian, and Haitian youth in New York City. *Race Ethnicity and Education*, *5*(1), 67–89.

Lorinc, J. (2003, September). Class struggle: Inside out deteriorating schools. *Toronto Life*, 98–99.

Louie, V. (2001). Parents' aspirations and investment: The role of social class in the educational experiences of 1.5 and second-generation Chinese Americans. *Harvard Educational Review*, *71*(3), 438–474.

Maynes, B. (2001). Educational programming for children living in poverty: Possibilities and challenges. In J. P. Protelli & R. P. Solomon (Eds.), *The erosion of democracy in education: From critique to possibilities* (pp. 269–296). Calgary: Detselig Enterprises.

McLaren, P. (1998). *Life in schools: An introduction to critical pedagogy in the foundations*. New York: Longman.

Milner, H. R. (2007). African American males in urban schools: No excuses—teach and empower. *Theory into Practice*, *46*(3), 239–246.

Milner, H. R. (2006). But good intentions are not enough: Theoretical and philosophical relevance of teaching students of color. In J. Landsman & C. W. Lewis (Eds.), *White teachers/Diverse classrooms: A guide to building inclusive schools, promoting high expectations, and eliminating racism* (pp. 79–90). Sterling, VA: Stylus.

Myles, J., & Feng, H. (2004). Changing colours: Spatial assimilation and new racial minority immigrants. *Canadian Journal of Sociology*, *29*(1), 1–35.

Ng, R. (1993). Racism, sexism, and nation building in Canada. In C. McCarthy & W. Crichlow (Eds.), *Race, identity and representation in education* (pp. 50–59). New York: Routledge.

Noguera, P. A. (2003). The trouble with Black boys: The role and influence of environmental and cultural factors on the academic performance of African American males. *Urban Education*, *38* (4), 431–459.

Odih, P. (2002). Mentors and role models: Masculinity and the educational "underachievement" of young Afro-Caribbean males. *Race, Ethnicity and Education*, *5*(1), 91–105.

Ontario Human Rights Commission. (2003) *Paying the price: The human cost of racial profiling*. Toronto: Author.

Portes, A., & MacLeod, D. (1999). Educating the second generation: Determinants of academic achievement among children of immigrants in the United States. *Journal of Ethnic and Migration Studies*, *25*(3), 373–396.

Preskill, S., & Jacobvitz, R. S. (2001) *Stories of teaching: A foundation for educational renewal*. Upper Saddle River, NJ: Merrill Prentice Hall.

Rong, X. L., & Brown, F. (2001). The effects of immigrant generation and ethnicity on educational attainment among young African and Caribbean Blacks in the United States. *Harvard Educational Review*, *21*(3), 536–565.

Silverman, D. (2000). Analyzing talk and text. In N. K. Denzin & Y. S. Lincoln Eds.), *Handbook of qualitative research* (pp. 821–834). Thousand Oaks, CA: Sage Publications.

Snyder, L. L. (2005). The question of "whose truth"?: The privileging of participant and researcher voices in qualitative research. In D. Pawluch, W. Shaffir, & C. Miall (Eds.). *Doing ethnography: Studying everyday life* (pp. 128–140). Toronto: Canadian Scholars' Press.

Taylor, D., & Dorsey-Gaines, C. (1998). *Growing up literate: Learning from inner-city families*. Portsmouth, NH: Heinemann.

Warren, S. (2005). Resilience and refusal: African-Caribbean young men's agency, school exclusion, and school-based mentoring programmes. *Race, Ethnicity and Education*, *8*(3), 243–259.

Waters, M. (1999). *Black identities: West Indian immigrant dreams and American realities*. Cambridge, MA: Harvard University Press.

Weis, L., & Fine, M. (2005). *Beyond silenced voices: Class, race and gender in United States schools*. Albany, NY: State University of New York Press.

Yosso, T. J. (2005). Whose culture has capital? A critical race theory discussion of community wealth. *Race, Ethnicity and Education*, *8*(1), 69–91.

Zhou, M. (1997). Growing up American: The challenge confronting immigrant children and children of immigrants. *Annual Review of Sociology*, *23*, 63–95.

Part III

Spirituality as Identity with Implications for Research and Teaching

6

Leadership and a Critical Spirit of Resistance

New Ways to Conceptualize Qualitative Research on Leadership and Spirituality

Michael E. Dantley, Miami University–Ohio

The field of educational research is replete with celebrated methodologies, procedures, and, in fact, a hegemony that prescribes what is considered to be genuine and valuable research. Scholars have engaged in numerous philosophical and ontological debates focusing on the efficacy of quantitative and qualitative research. In recent times, a discourse has emerged in research that salutes the poly-voiced character of researchers and the researched, celebrates culturally sensitive and relevant research, andendorses transformative and emancipatory research (see Mertens, 2005; Patton, 2002; Tillman, 2002). But all of these ways of positioning scholarship, it seems, are grounded in some sedimented notions of what "legitimate" research actually is. Within these prescriptions for scholarship is the idea that the purpose of research is to disclose some truth or truths and that these discovered truths are only legitimate if they have been unearthed through traditional or "accepted" forms of research methodology. Some positivist scholars argue that there exists what is called a grand narrative or some celebrated truth that purportedly essentializes the human condition—that research, it is argued, always either leads to or confirms its existence. The grand narrative is "a script that specifies and controls how social processes are carried out" (Stanley, 2007, p. 14). The grand narrative in education has been established through years of substantiating research. But what is more germane to this discussion is that not only does the grand narrative specifically locate or position education in a particular ontological and teleological discourse but also the research methods to sustain the tenets of the grand narrative are essentialized and cast in some reified notions of proper or vaunted research.

The purpose of this chapter is to call into serious question the hegemony perpetuating exigencies of the traditional and even the more progressive notions of qualitative research where spirituality is concerned; to use the work of Dillard (2000), Tillman (2002), Lather (1991), and Milner (2006) to deconstruct traditional notions of qualitative research; and to project a new way to conduct, consume, and evaluate the efficacy of qualitative research when exploring the mysteries of educational leadership and spirituality. This chapter will examine each of the scholars' contributions to the discourse on qualitative research and will then draw conclusions regarding how their conceptions of epistemology and research can contribute to the construction of an epistemological position that best grounds research in spirituality, leadership, and social justice. I begin with a brief discussion of my own ideological position concerning qualitative research and then explore the essence of the other scholars' contributions to this discourse starting with Milner's work.

Ideological Position on Qualitative Research

From my perspective, qualitative research in education and specifically educational leadership must be able to accomplish a number of objectives. First, qualitative research must be grounded in an effort to uncover the myriad ways schools and their leadership marginalize and perpetuate undemocratic practices particularly against poor and African American students. The goal of uncovering the instances of racist and classist malfeasance in educational leadership is not merely to expose such practices but rather to achieve a second project, which is to rectify the systemic procedures and practices that have allowed this disenfranchisement to exist in the first place. When perceived through this lens, scholarship and research then have a political agenda to fulfill. The knowledge gained through research exists not merely to fulfill the career aspirations of academicians but also to serve as a catalyst for the radical reconstruction of schools (Foster, 1986) and ultimately the broader society. This research calls into question the tenets of education's grand narrative and places the functions of the educational process in a broader social context.

To me, research must not only be individually sustaining and fulfill personal probative predilections but also be motivational for democratic communal reforms and societal reconstruction grounded in equity and a celebration of diversity. Given my personal assumptions about research, it seems only reasonable then that I would maintain that qualitative research that is conducted in order to highlight a social justice agenda ought to be theoretically grounded in progressive and critical conceptual frameworks. These theoretical positions actually serve three distinct purposes. First, they interrogate the asymmetrical relations of power that are inherent in educational institutions. Second, they denude whose interests are being served through the thinking and professional behaviors that are manifested through the act of education and schooling. Finally, they argue the point that a dominant culture that maintains cultural and political power continues to maintain its position of privilege through systemic mechanisms in the educational process that need to be critically engaged. These critical theoretical perspectives have served as guiding principles for my research on the intersections of critical spirituality, educational leadership, and social justice.

To examine critical spirituality, educational leadership and social justice forge a very unfamiliar landscape for those who review and make editorial decisions regarding research and scholarship. Because this scholarship "space" is so unfamiliar, it seems plausible—if not essential—for an epistemological position outside the traditional empirical ones to ground this genre of scholarship. My scholarship in spirituality and educational leadership is certainly best underpinned through qualitative research methodologies, but often even these can be somewhat limiting as they tend to secure the already-established grand narratives pervasive in educational research (see Stanley, 2007). Research in an area such as spirituality defies interrogation through traditional research methodologies due to its subjective or deeply internal focus. Thus, such research needs a methodology that understands the efficacy of deep personal reflectivity. That is, a research methodology is needed that accommodates the messiness of critical self-reflection within a broader communal context. The scholarship on leadership, spirituality, and social justice prizes the research participant's challenge to critique his or her personal behavior while concomitantly unraveling the myriad of ways his or her behavior has been founded upon a dominant, privileged, and marginalizing culture. At the same time, scholarship on educational leadership and spirituality asks of educational leaders especially to locate their work in a deeper context of meaning making and purpose as well as within a teleological and axiological backdrop. Thus this research needs to embrace an autobiographical process while at the same time contextualizing that process within a broader social and cultural critique. Milner's (2006) work, like mine, also situates research in a socially responsible space that celebrates self-reflections of the research participants.

Milner's Reflective Model

Milner (2006) has established a reflective model that he argues allows the researcher and the researched to internally consider their multiple selves. For Milner, these selves include the cultural, the racial, and the spiritual. Throughout his work, Milner argues that research in education must have an empowering nature. By this he means that research must bring to bear greater opportunities for the expression of individual and collective humanity by those individuals and communities being researched as well as the progressive dislodging of forms of systemic and blatant oppression that so often exist in the educational systems under investigation. For Milner, qualitative research must have an explicit social-justice agenda. He offers that the actual result of qualitative research is the unfolding of the oppressive ways in which schools operate against African American children especially and the creation of strategies to rid the educational system of such practices. In a quest to push empowering research into a position of efficacy and action, Milner has maintained that qualitative research, especially research focusing on African Americans, must include the spiritual dimension. He further maintains that there is great value in studying these spiritual elements and promotes the efficacy of the researcher recognizing and clearly and unapologetically articulating the influence his or her spiritual self has had on the research. While Milner advocates for the researcher openly locating his or her spiritual self, he nonetheless believes that often the spiritual dimension of the researcher's life is abandoned or strategically omitted in order for her or his work to be accepted by the powerbrokers in the academy. He writes, "Researchers, especially Black researchers, often find themselves forced to suppress their spiritual research. Moreover, data or findings in research studies that point succinctly to spirituality are often overlooked or trivialized because those in power often find such a focus immeasurable, irrelevant, or inappropriate to consider" (p. 372).

In order for spirituality, race, and culture to ground qualitative research as Milner suggests, he contends that a form of spiritual reflection should be included in the process. Further, Milner argues that spiritual reflection helps to catapult qualitative research to the realm of empowerment, that is, to a point where the research compels the reader to move to some form of definite action that propels a social-justice agenda. This introspection that Milner celebrates helps to define what research questions will be probed, what data will be collected and from whom, how data are analyzed, and what conclusions are ultimately drawn. Milner maintains that this kind of introspection is an essential piece to grappling with the realities of life in schools, especially the realities of African American life in the educational process. He argues, "Understanding others, then requires that researchers understand themselves as they attempt to understand others. Moreover, engaging in empowering research can be liberating in that those participating in research come to know new and expanded truths and realities that allow them to work for social justice" (p. 374).

What must be faced, however, is the potential disregard this type of research may experience by those who hold the power in the dissemination of scholarship in the academy. Milner says that the current literature in educational research is not without discussions on spirituality and leadership, spirituality and teaching, and spirituality and learning. He adds, "However, how one actually carries out such research is perhaps the most challenging task for researchers, especially because those in power often disregard and shun such discussions in conducting research" (p. 372). Stanley (2006) furthers Milner's argument by discussing the marginalization of research on faculty of color in predominantly White colleges and universities. Stanley offers four reasons for this phenomenon. She writes,

> First, they [scholars of color] represent a small number of overall full-time faculty; second, many scholars of color refrain from participating in such studies because their numbers are so small that they are easily identifiable; third, prior to the 1960's, they were not viewed as an important focus of research; and finally, these studies are often conducted by faculty of color, and many majority White faculty do not believe that these individuals can be objective when researching their own community. Therefore, research on the experiences of faculty of color is sometimes viewed by traditional, often White scholars as lacking in rigor. (p. 703)

I would argue that several marginalizing conditions help to cultivate what Milner presents as the disregard for research on spirituality. First, as he so aptly argues, those who are in power, including those who are the gurus in deconstructing and critiquing asymmetrical relations of power resident in traditional research methodologies, subscribe to disenfranchising epistemologies and research methodologies themselves in order to ground their own counterhegemonic scholarship. In other words, the texts that call into question hegemonic epistemologies utilize those same epistemologies in order to promote their counterhegemonic positions. There is a hierarchy of what counts as scientific research that even pervades antifoundational inquiry. Somehow, the dominant research paradigms that remain intact though resistant and counterhegemonic research are committed to interrogating the efficacy and relevance of these traditional paradigms. So, it seems as if the research "system" is almost a monolithic construct that defies interrogation and transformation.

The second condition—the idea that may cause this form of scholarship to be marginalized—is the fact that such research is couched in doubly castigated phenomena. Its focus is both non-White and nonlinear or non-rational. Research on the "exoticized" African American subject provides an interesting vehicle through which the dominant culture can explore the ways of the" mystical" Black culture. Such research offers ways to "deal with" African Americans—to understand how we function in order to get along with us better. However, when coupling exploration of the supposed African American mystique with the ethereal nature of spirituality, one has the recipe for trivialized, minimized, and often-disregarded research.

The goal of Milner's reflective model is to move the researcher, the researched, and the consumer of the research to a position of empowerment, that is, to a place of implementing some transformative action that furthers the social-justice agenda. What Milner's reflective process produces is a "coming to know" as Milner phrases it, that surfaces individual predispositions, sedimented values, and beliefs that color the ways in which the researcher may potentially go about the entire research project. I applaud the potential efficacy of Milner's reflective model. I am left, however with the following questions: Can research on spirituality and leadership ever be separated from political agendas? Is it possible to interrogate spirituality and educational leadership through a perspective that merely suggests how the powerful ways of embracing one's spirit can enhance the effectiveness of educational leaders' work? Or, does the alignment of spirituality with leadership automatically forge a political construction that is unavoidable?

I would agree that from Milner's perspective, research on Black life in schools and spirituality has within it a political-action agenda. This research is not banal and is not conducted to merely provide descriptions and definitions of the regimes of suggested truth that essentialize African American life. The whole point of reflecting, it seems as Milner would argue, is that research that is grounded in spiritual reflection has a greater propensity to inherently possess a strategy to right wrongs and eradicate practices of social injustice wherever the venue of the research happens to be. So essentially, research on spirituality, leadership, and social justice cannot afford the luxury of decontexualizing its analysis or conclusions or depoliticizing its implications. This kind of research has within it a call to action, a political agenda that moves the research beyond the confines of traditional academic review. Essentially, Milner's premise is that research on spirituality and African American life is grounded in what Cornel West (1999) calls the nihilism of Black life. And research in this area, therefore, must address pointedly these social and cultural realities with an eye to offer remedies and hope to the numbers of people caught in what seem to be hopeless realities. This is much like action research with this caveat: Milner's qualitative research firmly grounds its methodology and epistemology in some axiological space that calls for a response and action through its findings and conclusions. Culturally sensitive research (Tillman, 2002) may provide the initiating context for conducting and consuming Milner's action-oriented research on spirituality, leadership, and social justice.

Dillard's Work Grounds Tillman's Culturally Sensitive Research

Tillman (2002) has broadened notions of culturally sensitive research as established by Dillard (2000) and Kershaw (1990, 1992). What Tillman does is to locate within the traditional components of qualitative research the tenets of Dillard's extension of Hill Collins's (2000) notions of an endarkened feminist epistemology in qualitative research. Examining Dillard's assumptions regarding an endarkened feminist epistemology will establish the foundation for discussing Tillman's contributions of culturally sensitive research.

Cynthia Dillard's work alleges that the research in educational leadership may be prepared to "examine more culturally indigenous leadership in the academy" (Dillard, 2000, p. 661). Essentially, Dillard argues that an epistemology that accentuates the nuances of culture, especially the culture of African American women, must possess what she refers to as an instrumentality. That is, this epistemology must be able to produce or to do something that would lead to "transforming particular ways of knowing and producing knowledge" (p. 662). This is very similar to Milner's notions of research. There is an effectual standard of such scholarship that does more than pass the traditional, positivist benchmarks established by the academic hegemony. Dillard's instrumental research is designed to construct knowledge that would have no chance of being placed in the center of a discourse because it emanates from voices and experiences that have been traditionally disenfranchised and disregarded. This knowledge from instrumental research also comes through untraditional forms of research methodology and even from perhaps counterhegemonic ways of conducting research. Dillard's qualitative research methodology may privilege stories and narratives, may not necessarily be couched in one theoretical conceptualization, and may include more reliance on experience to create knowledge than even the accepted forms qualitative research celebrates. Instrumental research and epistemology have the added feature of moving the researcher, researched, and the consumer of the research to deliberative action that forwards a social justice agenda. Clearly, this feature differentiates Dillard and Milner's work from traditional notions of qualitative research.

Dillard deconstructs what she calls the metaphor of research as "recipe" and offers a more useful way of perceiving research through what she terms research as "responsibility." In the notion of research as recipe, Dillard locates the traditionally accepted detached relationship between the researcher and the researched. This relationship, according to the more traditional conceptions of research that fall within the category of recipe, offers that knowledge or truth is grounded in objectivity, and only through this detached association can truth ever be unearthed. However, Dillard's idea of research as responsibility is a way of perceiving the work of qualitative research in a more utilitarian sense. She comments, "Thus a more useful research metaphor arising from an endarkened feminist epistemology is research as a responsibility, answerable and obligated to the very persons and communities being engaged in the inquiry" (p. 663). What research as responsibility does, according to Dillard, is specifically place the narratives of African American women at the center of the endarkened feminist epistemology and also give voice to this epistemological frame through the polyphonia of African American women's voices. As a result of Dillard's musing, she has established six assumptions that ground her thinking about an endarkened feminist epistemology. Those assumptions, she maintains, have become essential to her work in addressing the asymmetrical relations of power with which all researchers, especially researchers of color, must contend in the academy. The second reason for Dillard's establishment of these assumptions is a much more practical one. Dillard argues that by accepting the tenets of an endarkened feminist epistemology, those who must sit in judgment on the efficacy and the validity of scholarship, when African American women are centered in the research, are then offered an opportunity to embrace an alternative way to think about and critique this research. The final rationale for the construction of Dillard's assumptions was to "challenge the all too prevalent idea that there is a unitary way to know, do and be in educational research endeavors" (p. 672). Dillard's endarkened feminist epistemology causes me to inquire if research in spirituality, leadership, and social justice must be couched in a paradigm of research as a responsibility. Much like Milner's premises about research, Dillard locates scholarship deeply and intimately in the intricacies of the lives and the social injustices lived everyday by the researched. Therefore there is a tremendous alignment between the research project and the alleviation of social ills that are not only unearthed through the research but also confirmed by it. From this foundation established by Dillard, Tillman develops her own notions of culturally sensitive research.

One of the purposes of Tillman's culturally sensitive research is to celebrate the multiple voices of researchers and the researched that are either on the margins of hegemony or are left out altogether. Tillman offers, "An argument for the use of culturally sensitive research approaches within the field of qualitative research and the framework presented here is based on the assumption that interpretive paradigms offer greater possibilities for the use of alternative frameworks, co-construction of multiple realities and experiences, and knowledge that can lead to improved educational opportunities for African Americans" (Tillman, 2002, p. 5).

Tillman maintains that the absence of culturally sensitive research approaches will continue a "void in what the larger research community knows and understands about the education of African Americans" (p. 4). What Tillman does is to extend both Dillard's and Kershaw's concepts of culturally sensitive and Afro-centric research by specifically locating culturally sensitive research methods in five areas: culturally congruent research methods, culturally specific knowledge, cultural resistance to theoretical dominance, culturally sensitive data interpretation, and culturally informed theory and practice. What Tillman's work does is to locate the culturally sensitive research framework within the structure and language to which the larger research community is accustomed. Doing so, undoubtedly, makes these antithetical notions potentially more palatable to the research community.

While I applaud each portion of Tillman's framework, for the purpose of this chapter, the third potion of her framework, cultural resistance to theoretical dominance, is most germane. Cultural resistance to theoretical dominance attempts to disclose the variety of ways traditional qualitative research paradigms "minimize, marginalize, subjugate, and exclude the multiple realities and knowledge bases of African Americans" (p. 6). What resonates with my own thinking is the imperative Tillman's concept demands to unmask the many ways traditional research silences the voices and minimizes the perspectives of the "Other." What I am left with to discover is the research epistemology or epistemologies and research methodologies that have yet to be created that will allow for the acrid critique and ultimate dismantling of marginalizing research methodologies. Tillman admits that she does not offer a prescription for culturally sensitive research approaches, and, given the purpose of her piece, it stands to reason that no prescription would be proffered. But when overlapping the poignant work of culturally sensitive research methodologies with a quest to create an epistemology and methodology that celebrate spirituality, I am left with a lingering inquiry and no resolutions.

I have, after considering Tillman's notions, raised the following questions:

- Is it best to align spirituality with a project of social justice?
- Does the alignment make spirituality more researchable?

These two questions are clearly founded upon the need to have research on spirituality, leadership, and social justice accepted in mainstream scholarship. While there is a practicality to these questions, they nonetheless tend to concretize the hegemonic intentions of the traditional educational research discourse. Those questions do nothing to trouble the assumptions or problematize the standards generally used to critique the validity of research. So even as I am attempting to find a counterhegemonic epistemological position, doing so is fraught with responding to the accepted ways of knowing and constructing knowledge. This dilemma resonates with what Dillard has experienced. She writes, "It is precisely at this point of representation, when the pressures to conform to the 'norms' of 'proper' scientific research are most difficult to resist, that I seek to recognize the cultural genesis and meanings of the lives of African American women researchers and to disrupt and unsettle the taken-for-granted notions surrounding the very goals and purposes of educational research" (Dillard, 2000, p. 665). Two other questions that might serve to locate my thinking in a more counterhegemonic or critical space are the following:

- Why is it that current research paradigms do not celebrate the notions of spirituality?
- Is it possible to create an endarkened spiritual epistemology that is particularly useful when examining African American spirituality and educational leadership?

The last question is fodder for more thinking. Whenever I have written about spirituality and leadership, questions always surface as to whether or not African American leaders solely embrace their spirits to impact their professional practice. I know this is not the case, but as Tillman, Dillard, and Milner have said, I know only too well the African American culture that richly colors my perspectives on spirituality and educational leadership. I am loath to essentialize and even to imply remotely that only African American people are attuned to their spirits. That would be essentialism at its finest. But I, as an African American, clearly embrace the socially just project of the equitable, quality education of African American children. That is not only significant to my research, but it is also my teleological commitment to improve the plight of my Black mothers and fathers and sisters and brothers through education, one of many venues that can ameliorate the current plight of people of color in this country. So I am left in yet another quandary. Two questions that affirm this quandary are raised: should spirituality be separated from a culture of Blackness, and are there any existing research epistemology and methodology that are particularly resonant with such a pointed civil rights agenda or can portions of extant epistemologies and methodologies be adapted to fulfill such a project?

The first two portions of the discussion have focused on notions of research from an African American perspective. What needs to be extracted from that discussion is the fact that traditional paradigms of research have minimized the contributions of African American–centered research. Traditional notions of research have marginalized the voices of the "Other," and when those voices have been included the hegemonic ways of research methodologies have tended to exoticize the experiences researched and reported about the Black community. Undoubtedly, the concept of spirituality may be "othered" as well. Because the whole idea of spirituality is shrouded in mystique and wonder, to apply traditional measures of scrutiny and review to spirituality research therefore seems to be self-defeating and almost pointless.

Part of the futility of the application of traditional models of research to spirituality, educational leadership, and social justice points to the nagging existence of a duality where all of this is concerned. There is the duality of linearity versus nonlinearity, what is celebrated in traditional research and what is not, an African American and non–African American voice and then, most germane to this chapter, there is the clear duality of spiritual versus nonspiritual epistemologies. Perhaps another perspective that should be considered in order to establish an epistemological position for research in spirituality is one that troubles the notions of dualities. Patti Lather's (1991) work will help in this regard.

Patty Lather: The Futility of Dualities and Emancipatory Discourse

Lather provides another perspective from which work in reconstructing epistemologies for research in spirituality, educational leadership, and social justice can be drawn. Her perspectives on postmodernism and discourses of emancipation are particularly intriguing and germane when applying them to deconstructing and creating new ways to ground educational research. Lather begins by arguing that one cannot conceptualize notions of postmodernism from a monolithic frame of reference, but one must pluralize these notions in order to create an adequate depiction of social realities. Such pluralization demands that essentialist binaries are also dismantled because, as Lather argues, postmodernisms preclude any such obligations to locate social reality simply in an either/or or both/and construction. She writes, "Philosophically speaking, the essence of the postmodern argument is that the dualisms which continue to dominate Western thought are inadequate for understanding a world of multiple causes and effects interacting in complex and non-linear ways, all of which are rooted in a limitless array of historical and cultural specificities" (p. 21).

When Lather examines Marxism in the light of a postmodern critique, for example, she argues that "it is also important to not position it [Marxism] as the 'demonic other' of postmodernism" (p. 23). The construction of dualities often demands that a demonic "Other" is also established that is clearly antithetical to that construct that has been hegemonically created. To unravel the notions that research in spirituality and education must conform to the structures of traditional research calls for researchers and the consumers of research to lift the demonic-other construction from this genre of research. What is essential is that research in spirituality, educational leadership, and social justice must be located in a space that celebrates the multiplicity of ways to perceive social realities. What is also necessary, where research in spirituality is concerned, is for there to be a broad acceptance of what Lather calls regimes of truth closely followed by the dismantling of restrictive binaries, "linear logics of Western rationality, and a foregrounding of ambiguity, openness and contingency" (p. 23).

In my mind, research in spirituality simply will not fit comfortably in the sedimented constructions of traditional research paradigms. My project here is not to demonize traditional paradigms either but to admit that within these reified frames of thinking are boundaries and restrictions that do unfortunate damage to the possibilities of unveiling multiple layers of reality that research in spirituality and education can bring. Lather, in providing more fodder for this line of thinking, cites Campioni and Gross (1983) who argue, "Why is it necessary to unify/solidify what may be fluid, diverse and changing, if not in order to block and control it? Diverse, changeable, strategic knowledges pose a potential threat that must be minimized—that of the incapacity of theory, of any theory to capture reality in its entirety or its essence" (Campioni and Gross, 1983, p. 24).

Concomitant with the deconstruction of dualities is the proposition that nothing is innocent. Lather discusses in a Foucaultian (1980) sense that within the construction of binaries is the presence of asymmetrical relations of power. Inherent in this way of thinking is that the dominant prong of the binary and the counterhegemonic side both are vying for dominance and contesting for the privilege to legitimately name social realities. So indeed, neither side is innocent. Both have an inherent political agenda at stake. Given these ideas, I am compelled to inquire if by searching for new or alternative epistemological grounding, I am desiring to position research in spirituality, educational leadership, and social justice as a prevailing paradigm alongside the others or if I am searching for ways in which this research that is so meaningful to me can be understood and accepted by the broader group of educational-research consumers. I am further questioning if, by pursuing this line of inquiry, I am not perpetuating a kind of binary opposition between the accepted epistemologies and research methodologies and what I am proposing. While I believe research in spirituality and education, especially what I have called critical spirituality, is emancipatory, I would hope that research in this area would not become oppressive to other ways of perceiving and pursuing research in educational leadership. In other words, research in spirituality ought to operate in a nonregulatory fashion; that is, research in this area must give glimpses, and multiple glimpses at that, of the ways spirituality is played out in educational leadership settings. If research in spirituality maintains dichotomous relationships with other fields of educational-leadership research, then this monolithic idea that one is accepted over the other continues. That does not appear to be a healthy way to inspire inquiry and the construction of knowledge. Again, Lather's work helps to bring some clarity to this issue when she looked at feminist perspectives on empowering research methodology.

Lather argues that to do feminist research simply means to put the social construction of gender at the center of inquiry. She further writes that feminist researchers organize their inquiry around the principle that gender deeply shapes the realities of our lives. She maintains that feminism is, among other things, a lens or perspective upon which questions for research are grounded. Rather than viewing spirituality as a dominating paradigm, I wonder if, like feminism, spirituality and leadership can be placed in the center of research from a perspective that argues that spirituality is a part of every person's being. In fact, research in spirituality may have this idea—the idea of the pervasiveness of spirituality in the creation of social realities—as its organizing principle. If one accepts Lather's proposition as a foundation, then that implies that one has had to relinquish the traditional notions of spirituality as religion and come to view it as another undeniable way to construct meaning in social realities. This will be difficult for some because of spirituality's ethereal nature. Indeed, spirituality itself inhabits a contested terrain. However, *if we can accept that gender, race, sexual orientation, and ability help to shape our constructions of reality, then it seems plausible that one's spirituality might also be located among this group of identifiers.* This is a tremendous leap of faith for some but perhaps, through the continuous writing about spirituality, it can become a part of the lenses through which social reality is viewed.

I have been taken with what Lather has called an ideological goal of feminist research. She writes, "The overt ideological goal of feminist research in the human sciences is to correct both the invisibility and distortion of female experience in ways relevant to ending women's unequal social position" (p. 71). It seems that Lather's construction of an ideological goal serves as another organizing principle for her research. Just as Lather's research has an ideological goal, Milner's research has one as well. His ideological goal was to move the researcher, the researched, and the consumer of the research to a position of empowerment and a position where transformative action that furthered a social justice agenda was the result. Dillard's ideological goal was to move research to a position of responsibility where the researcher became obligated to the very persons whose plight had been investigated, especially that of African American women.

As Lather and others situate the specific element of their research in the center of understanding the social order, it seems that doing the same with spirituality would suggest that spirituality is a basic organizing principle that "shapes/mediates the concrete conditions of our lives" (p. 71). So then spirituality, like feminism, would become a lens that would "bring into focus particular questions" (p. 71). Lather's work, as has that of the other scholars mentioned, compels me to ask questions regarding research in spirituality, leadership, and social justice. Lather writes about the efficacy of praxis-oriented research and argues that, in praxis-oriented inquiry, "reciprocally educative process is more important than product as empowering methods contribute to consciousness-raising and transformative social action" (p. 72). Undoubtedly, my hope is that research in leadership, spirituality, and social justice will be more than a way to produce scholarship. I would hope that such research would contribute to consciousness raising that would lead to transformative social action. The reciprocity in the educative process is another significant piece in doing such research. The whole scheme of learning, consciousness raising, and social-action transformation is the ultimate goal of research in spirituality, leadership, and social justice. So the question of weighing the residuals of the research over the product of the scholarship is a question that must be embraced. Another question that emanates from Lather's feminist perspectives on empowering research methodologies is what is the ideological goal of spirituality? What will research in spirituality correct? What unequal social positions will end through the research in spirituality? Lather further demands a major alignment between the research that is performed and a specific theoretical grounding. She writes,

> Too often we who do empirical research in the name of emancipatory politics fail to connect how we do research to our theoretical and political commitments. Yet if critical inquirers are to develop a praxis of the present we must practice in our empirical endeavors what we preach in our theoretical formulations. Research which encourages self and social understanding and change-enhancing action on the part of developing progressive groups

> requires research designs that allow us as researchers to reflect on how our value commitments insert themselves into our empirical work. Our own frameworks of understanding need to be critically examined as we look for the tensions and contradictions they might entail. (p. 80)

Lather's position on the alignment between research and theoretical grounding forces me to grapple with the question of what theoretical positioning grounds work in spirituality, educational leadership, and social justice. I am further compelled to discern if my personal value commitments are monogamous with this theoretical grounding and the resulting research.

Finally, Lather writes about self-critique and student resistance to liberatory curriculum. She spends time delineating notions regarding the politics of knowing and being known. She argues about the inadequacy of the self to be fully able to offer an auto-critique of validity and substance. Essentially, there is an underlying political essence beneath the questions we raise, the research we engage, and the theoretical dispositions we embrace. In understanding student resistance to liberatory curriculum and pedagogy, Lather wonders if therein lies the best form of self-critique. Deconstructing is a political event. It is not a valueless project that is totally empirically or conceptually grounded or epistemologically neutral. To deconstruct notions of leadership, spirituality, and social justice—that is, to formulate reasons why such ideas may be resisted and misconstrued by those who read research in this area—may be an essential avenue of self-critique that will help to couch this kind of research in a much more palatable context. Lather writes,

> To deconstruct the desire that shapes a particular act of enframing is to probe the libidinal investment in form and content of the author-text relationship. It is to mark the belief that our discourse is the meaning of our longing. Deconstruction agitates on behalf of the exuberance of life against a too-avid fixing and freezing of things. As I look at my own empirical research into student resistance to liberatory curriculum, deconstruction helps me to frame such questions as the following . . . Did I encourage ambivalence, ambiguity, and multiplicity, or did I impose order and structure? What elements of legislation and prescription underly my efforts? How have I policed the boundaries of what can be imagined? (pp. 83–84)

The most poignant question that comes from engaging Lather's work however is, "What is the fierce interest in proving the relevance of intellectual work?" (p. 84). What a powerful question. Grappling with this question will frame the substance of the conclusion of this chapter.

Conclusion

For some reason, perhaps birthed from ideas delineated in the beginning of this chapter, I have felt obligated to develop a cogent argument for the inclusion of research in spirituality in the celebrated canons of educational-leadership research. The fierce interest of proving the relevance of this intellectual work to me is actually that it will serve as an impetus to motivate the radical reconstruction of the social order in schools as well as the broader society. But if this research is left out of the major canons of educational-leadership research and scholarship, and if it is scrutinized through a lens that is bereft of the political underpinnings that ground this work, then the chances to transform through the traditional channels of research distribution become more and more limited. And yet, there is a revolutionary resistance kindling inside me that says this scholarship might be more accepted in the progressive, left-of-center venues that thrive on avant-garde scholarship. The questions scholarship for this chapter has birthed, however, far outweigh my original intent to forge a new epistemological grounding within which to locate spirituality research. The questions are formidable and actually serve as a phenomenal breeding ground for further conceptual scholarship on spirituality, leadership, and social justice.

Devoid of conclusions, I partially feel that the chapter has failed to meet a needed challenge. However, it is easy to essentialize and to enframe a text that promotes a paradigm of personal choice. It is much more meaningful and intellectually stimulating, however, to ask a foundational question and, through the process of answering the question, find the nascent of far more questions than answers. That has been the case with this chapter.

I do conclude that the scholarship in spirituality, leadership, and social justice must be written in a resistant voice. The scholarship must continue to trouble hegemonic notions of schools and their leadership. It must call scholars and practitioners to embrace introspection that will move the research, researcher, and the researched to the realms of empowerment. Research on spirituality must become research as responsibility and promote notions of cultural sensitivity, and, finally, it must grapple with many of the same questions that ground research in critical and poststructural frames. Though scholarship in spirituality and leadership has been flourishing, it is still a contested space that has a multitude of questions and challenges that have yet to be embraced.

References

Campioni, M., & Gross, E. (1983). Love's labours lost: Marxism and feminism. In S. Allen and P. Patton (Eds.), *Beyond Marxism: Interventions after Marx* (pp. 113–141). Leichardt, Australia: Intervention Publications.

Collins, P. H. (2000). *Black feminist thought: Knowledge, consciousness, and the politics of empowerment.* New York: Routledge.

Dillard, C. B. (2000). The substance of things hoped for, the evidence of things not seen: Examining an endarkened feminist epistemology in educational research and leadership. *Qualitative Studies in Education, 13*, 661–681.

Foster, W. (1986). *Paradigms and promises: New approaches to educational administration.* Buffalo, NY: Prometheus Books.

Foucault, M. (1980). *Power/knowledge: Selected interviews and other writings, 1972–1977.* New York: Pantheon.

Kershaw, T. (1990). The emerging paradigm in Black studies. In T. Anderson (Ed.), *Black studies: Theory, method, and cultural perspectives* (pp. 16–24). Pullman, WA: Washington State University Press.

Kershaw, T. (1992). Afrocentrism and the Afrocentric method. *The Western Journal of Black Studies, 16*(3), 160–168.

Lather, P. (1991). *Getting smart: Feminist research and pedagogy with/in the postmodern.* New York: Routledge.

Mertens, D. M. (2005). *Research and evaluation in education and psychology: Integrating diversity with quantitative, qualitative, and mixed methods.* Thousand Oaks, CA: Sage.

Milner, H. R. (2006). Culture, race and spirit: A reflective model for the study of African Americans. *International Journal of Qualitative Studies in Education, 19*, 367–385.

Patton, M. Q. (2002). *Qualitative research & evaluation methods.* Thousand Oaks, CA: Sage.

Stanley, C. A. (2006). Coloring the academic landscape: Faculty of color breaking the silence in predominantly White colleges and universities. *American Educational Research Journal, 43*, 701–736.

Stanley, C. A. (2007). When counter narratives meet master narratives in the journal editorial review process. *Educational Researcher, 36*(1), 14–24.

Tillman, L. C. (2002). Culturally sensitive research approaches: An African American perspective. *Educational Researcher, 31*, 3–12.

West, C. (1999). *The Cornel West reader.* New York: Civitas Books.

7

Awakening the Spirit

Teaching, Learning, and Living Holistically

Stephen D. Hancock, University of North Carolina–Charlotte

Our present era can be characterized as an "age of hedonism" that has produced a society of people who are focused on meaningless and secular fulfillment. This age has created a cycle of behavior where "the most popular response to the lack of meaning and the emptiness of life seems to be one of a highly intensified personal hedonism: an orgy of individual gratification in the form of consumerism; heavy reliance on sex, drugs, and music for release and distraction; and a never-ending pursuit of still greater heights of pleasure" (Purpel, 1989, p. 23).

We are in an age when the television has become a trough for the mindless appetites of excess, immorality, and self-delusion; when the military stands on the brink of impotency, while special-education classes swell with intelligent young men who are destined to fill the prisons; when the poor and oppressed are blamed for their condition, while the wealthy and powerful divorce themselves from responsibility; and when "I" is more important than "we," and "us" no longer rings with humanity. It is an age of emotional fragmentation, where our personalities are internally and externally disconnected and where wholeness is unwittingly traded for individual capital and myopic defenses. Our present moment in history marks a point where intellectual peace is substituted with hedonistic escapades and critical thinking is disregarded so as not to acknowledge the anguish of others. It is indeed a time of spiritual unconsciousness where even the attacks of September 11 could not revive the spiritual complacency, narcissism, and profligacy endemic in American society.

If, in fact, schools are microcosms of society, then we must logically conclude that the U.S. educational system is spiritually unconscious or a conscious contributor to its own demise. While the notion of self-denigration might be unconscionable to most, the idea serves only to put into perspective the current moral and spiritual crisis in American classrooms. School districts throughout the United States are plagued with high drop out rates, violence, inappropriate student-teacher relationships, mass killings, rampant academic failure, epidemic suspension and expulsion rates particularly among minority males, inequitable funding, and a lack of cultural and human connection. The condition of the educational systems in many U.S. cities and suburbs is such that either there is a conscious and purposeful effort to denigrate minority, poor, and, consequently, all students or the moral and spiritual consciousness in school policy, funding, teacher-student relationships, and curricular practices is incapacitated.

While inner-city and a growing number of rural and suburban schools are experiencing educational challenges, there is a sense that the problems of "those" schools do not impact the "good" schools, and thus these problems are relegated to "those" students, families, and communities. For years this segregationist and secular mindset has run rampant in American society and has quieted the voice of the spirit into a state of unconsciousness. Miller (2000) asserts that the dominance of secular mindsets and lifestyles has led to a repression of the spiritual life. It is also this disconnection from the spiritual life that enables a society of people to live in emotionally and physically segregated spaces. Unless people from all walks of life—those in power or subject to authority and those of diverse religions and worldviews—find a way to *awaken the spirit* to become rededicated to teaching, living, and learning holistically, we will continue to be subject to the conflict that is rampant in our schools and communities.

The full explanation of the spiritual, emotional, educational, and social dilemma in our schools is beyond this chapter; however, one does not have to search hard to read of the epidemic failure, the rapid resegregation of American schools, increasing dropout rates, inequitable funding, and dysfunctional relationships among students and teachers to come to the conclusion and confess that many American schools are in crisis. When one experiences the anxiety of elementary-school students who are forced to sit for two hours to take a test that they were not prepared for, witnesses students trying to learn in schools with unqualified teachers, sees the damaging results of oppressive and inappropriate curricular practices, or watches the light of learning "sucked out" of a generation of children who are culturally, economically, and academically marginalized, then one may be able to understand that the word "crisis" pointedly describes the state of many American schools. The broad stroke that paints the American school system in a spiritual crisis is based on the concept that spiritual crisis anywhere in American schools is a threat to spiritual progress everywhere in American schools.

Yet, in spite of the spiritual slumber and academic crisis in our schools, there is a voice in education that desires to shout against social and academic injustice, to listen effectively, to defy the monolithic perspective of school curricular, to nurture a sense of awe, and to care beyond safety. It is a voice that wishes to be known in an educational arena that has no capacity for it to be heard. It is the voice of the spirit. Palmer (1993) put it this way: "Conventional education neglects the inner reality of teachers and students for the sake of a reality out there, the heart of knowing self . . . is never given a chance to be known" (p. 35).

There is an inner voice that lives deep within the hearts of educators, and it yearns to sing of humanity, wholeness, and intellectual peace despite the spiritual crisis of American schools. Admittedly, humanity, wholeness, and intellectual peace might be seen by supporters of high standards and accountability as soft, slippery, and nonacademic. However, unless the educational community embraces spiritual tenants found in understanding self and others holistically, acting humanely and in the best interest of all students, and knowing the peace that comes by accepting the reality of a purpose greater than egocentric ambitions, there will be little progress in awakening the spirit. While I do believe there is a place for high standards, testing, and accountability in education, effective teaching and learning in the twenty-first century must embrace at least the fundamental pillars of humanity, wholeness, and intellectual peace in order to cultivate dynamic teachers and learners.

In an effort to make sense of how to awaken the spirit, this chapter will first grapple with an explanation concerning the definition and relationship between morals and spirituality. Second, the chapter will discuss a personal journey into the concepts of humanity, wholeness, and intellectual peace and their impacts on teaching, learning, and living in spiritual consciousness. Third, the chapter will address reflective practice as a fundamental activity that must be nurtured by educators and policy makers in order to usher in a spiritual awakening in American schools. Finally, the chapter will share a story of hope in learning, living, and teaching in spiritual consciousness.

Moral and Spiritual Consciousness

> There is something within man that cannot be reduced to chemical and biological terms, for man is more than a tiny vagary of whirling electrons . . . man is a being of spirit.
>
> —Dr. Martin L. King, Jr., *The Measure of a Man*

It is the "something within man" that King (1988) speaks of that seems to have been lulled to sleep by our current hedonistic age. This "something" refers to man's spiritual consciousness or the innate capacity to imagine, act, think, and commune beyond the physical realm. While a moral and spiritual conscious is present in every person and in various degrees, when the spirit is fully awakened it has the potential to alter ways of knowing and worldviews. However, because spirituality escapes rationality, and morals are sometimes inconvenient to our temporal constraints, they (spirituality and morality) are often dismissed as inconsequential and irrelevant. Nonetheless, man cannot ignore moral responsibilities any more than the body can divorce itself from spiritual yearnings. Therefore, matters of the spirit must be contended with, and educators must ask the following questions: What is meant by spirituality? What is moral behavior? And what does it mean to be a spiritually conscious educator?

Grappling with the relationship between spirituality and moral principles in an attempt to describe a clear and concise definition is not the goal. Rather, the goal is to provide a perspective that allows one to embrace notions of moral and spiritual consciousness as interdependent constructs. Spirituality is often associated with a concern for things higher than the realm of man where the spiritual focus is on sacred and eternally good forces (Palmer, 1993, 1998). While one can argue the case for good and evil notions of spirituality, this chapter is grounded in an idea of spirituality as the essence of life and the source of moral behavior.

The DK Illustrated Oxford Dictionary (1998) defines "spiritual" as a concern with the essence of life and a sensitivity to the intangible and sacred and attentiveness to matters of the soul. Attention to the essence of life and the intangible realm of being suggests that spirituality nurtures a sense of awe for something greater than human beings. Palmer (2003) describes spirituality as "the eternal human yearning to be connected with something larger than our own egos" (p. 377). Palmer's expression of spirituality promotes an attraction and identification to a rational or intelligent being greater than the mind and will of man. The affinity that man has toward God moves man to a pathway that demands the spirit to act in ways that lead to greater spiritual development. In 1995, the American Counseling

Association (ACA) described spirituality as "a capacity and tendency that is innate and unique to all persons. It moves the individual toward knowledge, love, meaning, hope, transcendence, connectedness, and compassion" (as cited in MacDonald, 2004, p. 294).

Describing spirituality as an innate capacity connects the intangible realm of being to the tangible actions of human behavior. Therefore, the ethereal realm of the spirit nurtures a space for the growth of moral principles. Spirituality, or spiritual consciousness, then, is where moral principles originate. Spirituality is also a place where the intangible realm of being, thinking, and knowing intersect with the practical behaviors or actions of man. Consequently, if one is to be spiritual and nurture an affinity toward God, there must be a move to behave or act in knowledge, love, purpose, hope, and compassion.

Fenstermacher (1990) defined moral behavior as "human action undertaken in regard to other human beings. Thus, matters of what is fair, right, just, and virtuous are always present" (p. 113). Moral principles then are concerned with human conduct and are often grounded in a religious context. Though the boundaries of religion are transcended by spirituality, religion often acts as an accountability measure for moral behavior. For example, the nexus of a moral concept might be found in Christian spirituality and thus express moral behavior rooted in Christianity. Here the Christian religion acts as the point from which moral principles originate and are launched. In addition, Christianity as a religious choice acts as a lens to help navigate the limitless journey into spiritual consciousness. Kesson (1994) suggested that morality, or spirituality without religion, is similar to traveling uncharted terrain where one is vulnerable to beliefs, images, and energies that can be unsettling and even dangerous. While spirituality is grounded in some form of religion, it is also able to transcend religious dogma to create a pathway to wholeness, intellectual peace, and humanity.

Given that spirituality is a part of most cultures, it is necessary that universal principles are employed to develop pathways to spirituality. While personally constructed spirituality is manifested in myriad forms (Aponte, 1998), religious and secular constructs are the two cannons of spirituality (MacDonald, 2004). Religious spirituality might be centered in Christianity, Judaism, Islam, Hinduism, Buddhism, Taoism, or any world- or culturally based religious group. Secular spirituality tends to have a focus on the temporal, the immediate, and the physical. It is concerned with the world we can see, and the physical pleasures and natural occurrences in the world (MacDonald, 2004). Despite the different foci of religious and secular spirituality, there are moral principals common to both forms of spiritual expression. I argue that any attempt to encourage moral and spiritual consciousness must appeal to all those who are involved in the educational process regardless of whether they aspire to religious, secular, or no spirituality. I believe that the principles of wholeness, humanity, and intellectual peace are comprehensive concepts for any spiritual disposition.

Wholeness, humanity, and intellectual peace are three dimensions of moral and spiritual consciousness that must be awakened in order to address the conditions in American schools. I have framed the concepts of wholeness, humanity, and intellectual peace as pillars of spirituality. Wholeness refers to a deep knowledge of the inner self that allows one to act in humility and compassion in an effort to combat personal fragmentation. Humanity is concerned with the moral interactions between and among people and how these interactions free or oppress the human spirit. Having peace with the truth about the inner life as related to others, God, and the environment characterizes the essence of intellectual peace. Therefore, spiritual consciousness requires that one is aware of the inner self in relationship to others, is awakened to the awe and wonder of God, and acts in a manner congruent with inner convictions and beliefs.

Education and Wholeness, Humanity, and Intellectual Peace

Compulsory education was initiated to improve the conditions caused by the Industrial Revolution, to maintain an educated workforce, to open pathways of upward mobility for all citizens, and to become the primary institution for the enculturation of American citizens (Fienberg, 1990). During this time, teaching was seen as a moral activity (Fenstermacher, 1990). The professional responsibility of teachers required that they teach the academic course of study and, more importantly, act as moral models. At the present time, in large scale, education is primarily interested in the transfer of technical information. Unfortunately, attention to matters of wholeness, humanity, and intellectual peace is not part of any priority. While school districts in the United States are right to be concerned with producing students who are prepared to enter the workforce as productive and educated citizens, the idea of nurturing the spirit should not be seen as another burden required of educators. Rather, in an effort to create holistic educational experiences educators must act to awaken the spiritual consciousness as a response to their professional responsibilities.

Wholeness

Wholeness in this endeavor encompasses the notion of personal harmony that enables one to combat spiritual fragmentation. It has much to do with the soul and is a process where the physical, mental, emotional, and social selves discover purpose, congruence, courage, and a critical understanding of the interconnectivity of life. Wholeness seeks to connect self to self in order to combat the fragmentation of thought and being. Teachers are able to live in fullness when there is a conscious effort to seek wholeness. Through wholeness, the departmentalization in ways of thinking, knowing, and acting gives way to a comprehensive awareness of self. The deep knowledge of the inner self acts as the first pillar to understanding the self as having one spirit and one purpose. Wholeness longs for a place of congruency where the total person is realized in his or her most excellent form (Hancock, 2003).

In the classroom, wholeness gives way to a holistic identity that enables educators to act in ways compatible with inner convictions about the art and act of teaching, the culture and intelligence of students, and their processes and styles of learning. Educators who function from a state of wholeness are conscious of how personal experiences and beliefs impact perceptions of students, self, and the curriculum. As a result, wholeness creates a capacity for teachers to encompass culturally and developmentally appropriate teaching methods. To teach holistically is to be responsible and capable of operating with personal and professional congruence as well as to know and critically reflect on motives, impulses, and belief systems (Green, 1988). The act of critically reflecting on our ways of knowing and shortcomings should move educators into a place where they become internally competent and whole.

Teachers who operate in wholeness are able to abandon pretentious acts of fulfillment and recognize dormant areas and problematic beliefs and behaviors in their lives (Palmer, 2003). Many teachers often say that they are fair, loving, caring, and respectful only to discover through critical reflection that they treat students with contempt, sarcasm, and condescension. The fragmentation found in this behavior speaks to unexamined realities that influence teacher behavior. Wholeness requires that teachers confess their beliefs or issues concerning students, the curriculum, and teaching. True confession of problems neither frees teachers to judge nor cages them in condemnation. Rather, it creates a space of humility, empathy, and compassion where teachers are able to know and act holistically. The acknowledgment of their voids enables educators to reflect on their shortcomings in an effort to gain harmony with their thoughts and actions. Teachers who confess and acknowledge personal voids are better able to combat fragmentation in their personality and move toward a state of wholeness.

The process of becoming whole is just that—a process. It is not an attainable goal or firm stepping stone to another spiritual dimension. Wholeness is a spiritual pillar that enables teachers to understand the whole self in an effort to educate the whole child. The notion of understanding the genuine self should nurture compassion, humility, and positive actions towards others. For teachers, wholeness functions to bring peace and congruence to inner convictions and external behaviors as it relates to students and the curriculum.

Humanity

While spirituality is deeply personal, it is also intricately connected to interpersonal interactions. The relationships between and among humans are key factors in our spiritual development. Thus humanity is concerned with the moral interactions between and among people and how these interactions free or oppress the human spirit. When acting on such moral principles as compassion, peace, cooperation, honesty, and respect one acts in a humane posture poised to free the spirit of self and others. Hence, the notion of humanity dictates that man is required to embrace positive and responsive interactions between and among others.

Palmer (1993) describes education similarly as "being drawn into personal responsiveness and accountability to each other and the world of which we are a part" (p. 15). If teachers are to be effective, accountability and responsiveness must be grounded in such moral principles as respect, love, peace, honesty, care, and compassion. Unfortunately, in classrooms across the United States, these moral principles are ingenuously practiced, leaving students to learn in dehumanizing environments with culturally insensitive curricula. Freire (1999) describes dehumanization as an oppressive force that seeks to denigrate others in an attempt to build up the oppressor. He contends that dehumanization is a product of moral and spiritual bankruptcy. In education, dehumanizing practices marginalize students through a series of oppressive teaching methods, curricular programs, and discipline policies. If educators are to humanely teach and learn, they must free themselves from oppressive curricula, ineffective methods, and dysfunctional relationships in order to become moral models and activists for students and others.

Fenstermacher (1990) stressed that teachers should model morality in order to be imitated by and influential with their students. He describes a classroom where teachers act justly while showing compassion, respect, care, and awareness of students' cultures (Fenstermacher, 1990). Genuine care for students and their community places students and teachers in a relationship of respect and influence. Fenstermacher (1990) presupposed that when a teacher acts with a humane or inhumane disposition, his moral character is on display. Therefore, "every response to a question, every assignment handed out, every discussion on issues, every resolution of a dispute, every grade given to a student carries with it the moral character of the teacher" (p. 134). When students and teachers are in a relationship characterized by a fundamental and authentic love and respect for humanity, it may increase the likelihood that students will want to learn. In short, "the relation of teacher and student must be deeply human for real learning to occur" (Palmer, 2003, p. 380). The concept of humanity as it relates to education is concerned with the moral interactions between and among teachers and students and how these interactions liberate students and teachers to learn in peace and reverence.

Intellectual Peace

Everyone is granted the capacity to think, to reason, and to understand. This capacity is intrinsically tied to the moral principles of humanity. Intellect or moral reasoning provides a capacity to know right from wrong and good from bad. Intellectual peace enables one to know when to reason and when to accept the unreasonable, when to process meaning and when to be processed by meaning, and when to be known and when to know (Palmer, 1993). Intellectual peace, therefore, is not influenced by IQ tests, factual knowledge, or the number of degrees attained; rather, intellectual peace is the harmony one finds in the knowledge of and a relationship with truth where truth is allowed to illuminate reasoning, understandings, beliefs, and motives (Palmer, 1993).

It was intellectual peace that was robbed from Adam and Eve as their knowledge of truth, self, and God was thrown into chaos with a single act of hedonism. Even today, there is a continual struggle with submitting to truth, understanding the self, and reverencing the awe and wonder of the Creator. Spirituality couched in Christianity understands truth to be a living incarnation in the person of Jesus Christ. Thus, if a Christian is to possess intellectual peace, he or she must submit to truth and allow it to build and dismantle, to reveal and revile, and to encourage and caution in an effort to gain knowledge of self and God. Intellectual peace nurtures a capacity to know truth, self, and God in a harmonious symphony. Having peace with the truth about the inner life as related to others, God, and the environment characterizes the essence of intellectual peace.

The initial movement toward peace is often paved with chaos, as we have to work past dogmatic assumptions and cemented beliefs. In a world where truth is relative and situational, it is admittedly difficult to hold truth as both absolute and liberating. Yet, without entering into relationship with truth and submitting to what is known as innately good and righteous, man may never be able to know the benefits of being at peace with self, the truth, and God. It is the challenge of humanity to submit to truth in order to obtain the "peace of God, which transcends all understanding" (Phil. 4:7). This peace promotes purpose and a reverence for the sanctity of life and for the awe and wonder of God.

As educators begin to nurture personal wholeness as it relates to humanity, they also are able to journey into truth. It is in truth that educators find their purpose and are able to nurture students in fulfilling their goals. Truth acts to judge personal notions and liberate teachers into a peace characterized by a growing knowledge of God, self, others, and their environment. The peace in knowing and growing personally and professionally requires that teachers submit to truth and dismiss the influence of control and superiority in the classroom.

Unfortunately, in education "we find it safer to seek facts that keep us in power rather than truths that require us to submit" (Palmer, 1993, p. 40). Educators have traded in intellectual peace and now operate from positions of oppressive power and intellectual chaos. If teachers, principals, and other school personnel began the process of finding intellectual peace, it is possible that the negative policies and practices that impact marginalized populations of students might be revealed. Possessing intellectual peace will enable educators the capacity to look past stereotypes, self-interests, and complacency and act on truth, compassion, and strength to better educate students and the educational community. Intellectual peace is the capacity of an individual to know and accept self as related to God, truth, and others (Palmer, 1993).

Reflective Practice and Pathways to the Spirit

Awakening a spiritual yearning that moves people to compassion, knowledge, love, meaning, and hope requires a systematic and comprehensive approach. In an effort to commune with something greater than the will and mind of man, teachers must be willing to examine their personal and professional selves. Educators must move toward spiritual awareness through active participation in reflective practice. Reflection represents a basic practice that should not be negotiated in the lives of educators. While educators may be tentative about committing to a critical reflective process, they must acknowledge the need and understand the ultimate benefits. Reflective practice as a fundamental process will help educators awaken from a hedonistic slumber into spiritual consciousness.

Reflective practice as a spiritual process requires teachers to confess and systematically explore their attitudes, beliefs, and values in an effort to nurture congruence with internal convictions and external actions. As educators reflect they must thoroughly respond to critical questions that are designed to awaken a moral and spiritual disposition. Questions might include the following:

1. What do I know about self, students, and teaching?
2. How have I come to know about self and others, and from where has this knowledge come?
3. How am I known by students, colleagues, administration, community, family, self, and so on?
4. What does my behavior reveal about my beliefs and spirituality?
5. How does my knowledge or lack of knowledge of self hamper what is possible in my personal life, professional life, curriculum, and instruction?
6. Why do I teach?
7. What motivates me to remain in the classroom?
8. How does spirituality influence my teaching practices?
9. How do I navigate fears, prejudices, ignorance, guilt, and control in the classroom and in my life?
10. What is my purpose in the lives of my students, colleagues, and school community?

The aforementioned questions can initiate a process of reflective practice. Reflective practice can start a process of awareness where teachers grow in capacity to embrace wholeness and humanity in an effort to abandon delusional notions of truth. Also, reflective practice can combat the fragmentation caused by moral and spiritual unconsciousness. For instance, reflective practice can assist educators in viewing themselves in wholeness, understanding that they are one entity with one spirit. The attention teachers give to the numerous hats they wear often diverts energies from their emotional and spiritual needs. Unfortunately, this fragmentation acts as a barrier designed to shut out teachers' ability to hear their spiritual voice. A benefit of reflective practice is that it nurtures a critical knowledge of the whole self that enables educators to know themselves fully and operate holistically, offering love, care, interest, and respect in whatever capacity they are found. Living holistically empowers educators to understand that their lives and the lives of their students are directly connected and mutually influential.

Another benefit of reflective practice empowers teachers to discover purpose. Critical reflection enables educators to nurture a spiritual purpose that will sustain them and their students in times of challenge. Purpose, in this instance, is not career oriented or task specific; it is a based on a passion to enhance the lives of others. Teachers who are able to look beyond grades, socioeconomic status, race, gender, academic mobility, and dispositions of students and nurture their wholeness and humanity are often spiritually conscious practitioners who understand that their purpose is to uplift and encourage each student.

One other benefit of reflective practice is that teachers become deeply aware of what they know and how they are known by others. The knowledge revealed in reflection is not just concerned with academic or skill-based knowledge but is more focused on what is known about the personal and spiritual self and how this knowledge impacts how others know us. Reflection can usher us into an intellectual peace that reveals strengths, challenges, and latent talents, intelligences, and gifts. In addition, reflection can also combat self-deception and deficit beliefs that fester in spiritual unconsciousness.

Purpel (1989) explained that "self-deception not only involves denial, fear, avoidance, and fragmentation, but it is also ultimately self-defeating" (p. 62). Educators who have not examined their lives operate in self deception. Teachers often say that they believe all children can learn but only teach, nurture, and support children who embrace their ways of teaching. In this instance, teachers unwittingly marginalize and dehumanize those students who possess forms of intelligence different from those targeted by their teaching methods. Consequently, these teachers are not able to act in the best interest of the academic, social, or spiritual lives of these students. Teachers who are not reflective about the inner self may remain in a spiritual slumber and may be unable to act on spiritual convictions. Self-deception undermines any attempt to act from convictions, beliefs, and compassion (Purpel, 1989).

A Story of Hope

It is no small feat and there is no one way to awaken the spiritual conscience. While I stress using the pathway of reflective practice, other practical opportunities for teachers to act from a spiritual angle are abundant but often ignored. Teachers may ignore these opportunities out of fear, frustration, and an elusive pressure from somewhere in the murky reality of educational policy. Yet, there are many instances where teachers act with a spiritual voice and despite fear, frustration, and educational pressures are able to embrace the wholeness, humanity, and peace in teaching and learning.

I now turn to a deeply personal story of hope in the face of pressure—a story where the spiritual conscience triumphed amid tremendous obstacles:

> I was a junior in high school, and, while I enjoyed learning, wanting to learn was not synonymous with going to school. It seemed to me that schooling was not designed to create learners but to produce masses of mediocre citizens poised to nurture the status quo. While I couldn't articulate my position, I pushed against this notion of schooling and found myself at odds with teachers and their idea of learning. Everything changed my junior year when I met Mrs. Baylor. Mrs. Baylor taught algebra, and I was a self-professed math phobic. But, instead of giving up on me, she tapped into a gift that I did not know I had. Though I struggled with quadratic equations and Pythagorean theories, Mrs. Baylor saw that my gift of teaching would be an asset to learning. So, I was often asked to work on a problem all week in order to present it to the class. While my grade point average remained a C– for the entire year, teaching the equations on the board nurtured a sense of confidence and aptitude in math. It was during final exam week, however, that Mrs. Baylor's greatest act of the spirit took place. In order to maintain a C in the course I had to pass the exam with a C. The next day we all waited for the results of our exam. As the class was dismissed for lunch, Mrs. Baylor stood at the door and gave each student a small card with their exam grade. When it was my turn, Mrs. Baylor motioned for me to step aside. When the room was emptied she told me that I had failed the test and would likely fail the class. Then against academic policy and traditional convention she said, "I know you know these equations so if you can correctly demonstrate them on the board I will award you the points." Though I was not shocked by her actions, I was stunned by her compassion and courage. I needed fourteen points to pass the test with a C. So, for thirty-five minutes, I demonstrated eight exam equations (for sixteen points) on the board while she watched intently. As other students began to file in from lunch, Mrs. Baylor hushed them, and they all sat and watched. When I finished she instructed me to stand aside while she reviewed each equation for the class. She corrected none. Then with a smile she said to the class and directly to me, "These are the correct answers to eight equations from your exam. If you answered your equations like this than you've passed this course." Despite the conventions of testing, the stigma of cheating, and the idea of being out of order, Mrs. Baylor stood firm in her convictions and beliefs about the wholeness of her students. She validated my kinesthetic learning style, my ability to present equations as a teacher, and my true capacity to do math. She could have easily failed me, but instead she nurtured my sense of hope in holistic education. That afternoon Mrs. Baylor gave me hope in schooling, but more importantly she showed me courage, conviction, and how to teach holistically, humanely, and with intellectual peace.

Looking at my exam situation from a secular or purely professional position would have warranted me receiving a failing grade. This failure would only be sanctioned because I did not pass a test—not because I was truly inept at understanding math. And even if I were completely inept at math, it is the teacher's responsibility to look holistically at the student and teach from a place of humanity, wholeness, and intellectual peace. Mrs. Baylor faced a spiritual crisis as she took on the responsibility to educate from a place of wholeness. Palmer (2003) asserts that "a spiritual crisis arises when we find ourselves in the grip of something larger than society's expectations or the ego's needs" (p. 377). The school system expected Mrs. Baylor to act in the best interest of the test to maintain what some might describe as academic integrity. While academic integrity has its place, human integrity in educational settings must be noted. Mrs. Baylor had a choice to either follow the voice of societal expectations or listen to her spirit. I am glad that she listened to her spirit and held to her convictions to educate for liberation and empowerment.

Palmer (2003) describes this type of teaching as that which enhances the human condition and sets the stage to launch social activism. Activism takes on many forms in the classroom, and Mrs. Baylor displayed activism with conviction. While neither Mrs. Baylor nor I had any idea that I would become an elementary-school teacher, her courage in the face of oppressive conditions sparked in me a renewed passion for art and the act of teaching. Mrs. Baylor found a way to navigate the demands of high-stakes testing, monolithic curricular methods, stoic teaching practices, and antiquated ideas of learning while at the same time teaching with integrity, compassion, and conviction. Throughout the year, she compelled her students to understand their abilities and challenges in an effort to chart personal successes through math. She used her power to nurture our learning through genuine interactions and engaging dialogue. Mrs. Baylor's reverence and connection to the sanctity of life fueled her compassion to nurture students beyond simply mathematics, as well as narrow ideas of intelligence and dehumanizing teaching practices. Her awareness and practice of love, respect, honesty, and responsibility enabled her to act with a spiritual conscience.

Hope can also be found in efforts like The Living Values Program that takes a holistic view of students and recognizes that peace, respect, love, tolerance, honesty, humility, cooperation, responsibility, happiness, freedom, simplicity, and unity are fundamental moral principles that must be present if quality education is going to take place (Arweck & Nesbitt, 2004). The hope of spiritual consciousness in education is realized on small scales in school districts throughout the United States. Teachers like Mrs. Baylor exist as refuges for students who are marginalized by oppressive and dehumanizing teaching practices. There are teachers who are actively combating the dehumanization found in schools as a result of high-stakes testing, insensitive curricular practice, and narrow ideas of intelligence.

Conclusion

In his book *The Measure of a Man*, Martin Luther King, Jr., (1988) framed wholeness, humanity, and intellectual peace as being the length, breadth, and height of life. He called length, breadth, and height the three dimensions of the spiritual life. King contended that the first dimension, length, is not concerned with how long one lives but how one develops their inner powers to become whole. The length of life or wholeness often breeds a people who might know themselves in isolation. King (1988) suggested that this knowledge is incomplete. If we are to know ourselves in wholeness, we must seek to know and be known by humanity (Palmer, 1993). In our classrooms, teachers and students must strive to know themselves in relation to others so as to understand their purpose and place. In fact, everyone has the responsibility to discover their purpose and to do it with all the strength and power they possess (King, 1988). This purpose should not be based on selfish gain but focused on enhancing the lives of all in the educational community.

The second dimension of humanity is referred to as the breadth of life. King (1988) stated that "an individual has not started living until he can rise above the narrow confines of his individualistic concerns to the broader concerns of humanity" (p. 42). Living humanely, therefore, dictates that we develop positive interactions and responses to others in order to fully know the breadth of our lives. Humanity, then, is "an inescapable network of mutuality" where we are an interdependent society (King, 1988, p. 48). Educators must prepare curricula that seek to combat the reality of dehumanization in our schools. Educators should not walk in the pretense of success when students are not nurtured in wholeness, humanity, and peace. Unless students, teachers, and the education endeavor are grounded in academic and spiritual success, humanity in its fullness is not realized. King (1988) believed that the spiritual bond of humanity was so strong that he confessed, "I can never be what I ought to be until you are what you ought to be" (p. 48). Teachers, administrators, and policy makers should adopt this perception of humanity and require schools to act and operate in the best interest of students, parents, and the entire educational community.

King (1988) says of the height of life or intellectual peace, the third dimension, that some people never reach this dimension. Consequently, their philosophies and spiritualities end at the glory and power of man. Therefore, if we are to nurture the spiritual we must reach beyond our wholeness and humanity into a place greater than our own will and mind. The height of life refers to a reverence for God where we must submit to a relationship with truth.

If we are to awaken the spirit, we must begin the journey by embracing the length, breadth, and height of life. As a society we must examine our lives while focusing on matters of wholeness, humanity, and intellectual peace. Teaching and learning must be viewed as a moral enterprise and thus require us all to examine our purpose, motives, and capacity to develop a spiritual conscience. Palmer (2003) demands that educators who choose to act from an unexamined life should not be allowed to impose their lives on students. Unfortunately, schools are full of teachers who live unexamined lives. Waking up our schools from a spiritual slumber requires a group of educators determined to live out the principles of wholeness, love and respect for others, and reverence for the awe and wonder of something larger than the will and mind of man. However, we have "become so involved in the things of this world that we are unconsciously carried away by the rushing tide of materialism which leaves us treading in the confused waters of secularism" (King, 1988, p. 50).

Yet there is always hope.

References

Aponte, H. (1998). Love, the spiritual wellspring of forgiveness: An example of spirituality in therapy. *Journal of Family Therapy, 20*, 37–38.

Arweck, E., & Nesbitt, E. (2004). Values education: The development and classroom use of an educational programme. *British Educational Research Journal, 30*(2), 245–261.

DK Illustrated Oxford Dictionary. (1998). New York: DK Publishing.

Fenstermacher, G. (1990). Some moral considerations on teaching as a profession. In J. Goodlad, R. Soder, & K. Sirotnik (Eds.), *The moral dimensions of teaching (pp. 130–151)*. San Francisco: Jossey-Bass.

Fienberg, W. (1990). The moral responsibility of public schools. In J. Goodlad, R. Soder, & K. Sirotnik (Eds.), *The moral dimensions of teaching (pp. 155–187)*. San Francisco: Jossey-Bass.

Freire, P. (1999). *Pedagogy of the oppressed*. New York: Continuum.

Green, M. (1988). *The dialectic of freedom*. New York: Teachers College Press.

Hancock, S. (2003). Balancing act: A reflective practice. In A. Green & L. Scott (Eds.), *Journey to the Ph.D.: How to navigate the process as African Americans (pp. 74–89)*. Sterling, VA: Stylus.

The Holy Bible New International Version. (1984). Grand Rapids, MI: Zondrevan Publishers.

Kesson, K. (1994). An introduction to the spiritual dimensions of curriculum. *Holistic Education Review, 7*(3), 2–6.

King, M. (1988). *The measure of a man*. Philadelphia: Fortress Press.

MacDonald, D. (2004). Collaborating with students' spirituality. *Professional School Counseling, 7*(5), 293–301.

Miller, J. (2000). *Education and the soul.* New York: State University of New York Press.

Palmer, P. (1993). *To know as we are known: Education as a spiritual journey.* San Francisco: Harper and Row.

Palmer, P. (1998). *The courage to teach: Exploring the inner landscape of a teacher's life.* San Francisco: Jossey-Bass.

Palmer, P. (2003). Teaching with heart and soul: Reflections on spirituality in teacher education. *Journal of Teacher Education, 54*(5), 376–385.

Purpel, D. (1989). *The moral & spiritual crisis in education: A curriculum for justice and compassion in education.* New York: Bergin & Garvey.

Part IV

Culture, Curriculum, and Identity with Implications for Teacher Education

8

Race, Narrative Inquiry, and Self-Study in Curriculum and Teacher Education

H. Richard Milner IV, Vanderbilt University

It is critical for teacher educators to examine their own practices because what they do, say, and model in the classroom have the potential to influence teachers and students in P through 12 classrooms. In this chapter, I use narrative inquiry and self-study to examine my own curricula decisions in teaching race in a course with mostly White teacher-education students. Race, of course, is socially and legally constructed (see, for instance, Ladson-Billings & Tate, 1995; Tate, 1997). It does, from my research and experience, involve the interpretation of an individual's skin color. Individuals from different biological groups are more the same than they are different. Studies that suggest that biology/genes are to blame for disparities, for instance, are faulty. These same analyses also suggest, however, that race is so ingrained in how we view each other that it cannot be overlooked or ignored (Bell, 1992; Ladson-Billings, 1998). Using racialized narratives in teacher-education courses can help circumvent resistance and disengagement among education students where race and racism are concerned. Such a focus is important to the education of students in all schools, and particularly to students in urban schools. Urban schools are often populated by students of color and students from lower social classes, while teachers are overwhelming White and middle class.

Much of the attention on the narrative in teacher education has been placed on students (the prospective and practicing teachers) enrolled in

This chapter has been reprinted/adapted from Milner, H. R. (2007). Race, narrative inquiry, and self-study in curriculum and teacher education. *Education and Urban Society, 39*(4), 584–609.

courses and in teacher-education programs. In this chapter, as an African American male teacher educator, I focus on the personal narrative and how my personal racialized story outside of the classroom seemed to influence my curriculum development and teaching in a course designed for teacher-education students. What was it that I, as an African American male teacher educator, brought into the learning environment of my college classroom that could enable or stifle the students' (preservice teachers') learning about race and racism in education? That is, how did my racialized experiences influence my curriculum development and teaching? Moreover, what roles can narratives and self-study play in students' opportunities to learn about race and racism in education? This research was a self-study in teacher education from which other teacher educators and teachers can learn.

Self-Study in Teacher Education

Self-study—teacher educators' systematic examination of their own practice to improve their work—is recognized as an important area of concentration in teacher education (Cochran-Smith, 2000; Hamilton & Pinnegar, 2000; Ladson-Billings, 1996). In fact, there is an entire special interest group within the American Educational Research Association that focuses its work succinctly on issues of self-study. However, Hamilton and Pinnegar (2000) expressed their concern that much of the teacher-education discourse is absent of teacher educators' voices and perspectives: "We do not hear the voices of those who create living educational theories" (p. 235) in teacher education. Dinkelman (2000) reminded us that self-study can result in teacher educators "knowing something important about [their] practice that [they] did not know before, something [they] only came to know about as a result of self study" (p. 7).

In many respects, self-study in teacher education has meaningful connections with action research that tends to have a clearer focus on P through 12–classroom inquiry. Action research, according to Rearick and Feldman (1999), "is to understand, then explain. The explanations are related to the desire to achieve greater clarity about the relationship between his or her inner state and action" (p. 335). As a teacher educator, Cochran-Smith (2000) used narrative and self-study to unlearn racism. In both action research and self-study, reflection is often used as a means for studying the self (see Dinkelman, 2000; Hatton & Smith, 1995). That is, individuals engage in deep levels of introspection in order to come to terms with both conscious and unconscious phenomena and experiences. The reflective process can shed light on situations that can help teacher educators reconceptualize their work. Indeed, to study the self is not an easy challenge. I agree with Hopper and Sanford (2004) that "our challenge as teacher educators . . . has been to examine our own assumptions about the value of the knowledge we offer and the ways in which we offer this knowledge" (p. 71). Coming to terms with what we know and how we attempt to share our knowledge requires us and our teacher-education students to come to terms with our knowledge as a foundation from which we can build. In addition to self-study, the narrative in teacher education has important implications for this research.

Narrative in Teacher Education

Narrative research has been used to shed light on the complex nature of teaching and learning (Kienholz, 2002; Pinnegar, 1996; Rushton, 2004; Valdez, Young, & Hicks, 2000). It has been used to explore cultural diversity in teacher education (Dome et al. 2005; McVee, 2004; Phillion & Connelly, 2004) as well as professional development in inservice teacher education (Elbaz-Luwisch, 2002; Schwarz, 2001). Alvine (2001) wrote, "Teacher educators have recognized the importance of the individual's lived experience as relevant to the development of what he or she will bring to the classroom. Thus, the life histories of teachers [and teacher educators] have come to be seen as grounded experience for knowledge of teaching" (p. 5).

McVee (2004) explained that "narratives of personal experiences need to be represented in teacher education courses in ways that demonstrate their [teachers and teacher educators'] dynamic, multiple viewpoints" (p. 897). Moreover, she suggested that "individuals can be transformed by stories and that stories can be a means for teachers to express beliefs about theory, practice, and curriculum" (p. 881). Using narratives in classes can help students—that is, preservice teachers—locate experiences that can guide their thinking and teaching.

Elbaz-Luwisch (2001) stressed that "we need to pay attention to our own stories as teachers [and teacher educators] if we are to be able to attend to the stories of pupils. This is the case at all levels of education" (p. 133). I attempt to discuss how I developed and transformed my curriculum with my students by modeling my curriculum and teaching so that they could consider narratives that could transform their own work with students in P through 12 schools. My proposition is that the teacher has the potential to transform the curriculum to such a degree that she or he *is* the curriculum (McCutcheon, 2002). The curriculum can be seen as a verb in this sense and not only as a noun (Pinar, 2001). To explain, the curriculum actually does something; if teachers are the curriculum, what they teach, how they live, what they model, what they say, and where they focus all have the potential to shape students' learning. Thus, the curriculum is a verb as well as a noun. It *does* something.

Students in the course seemed to move from a "prove to me that race and diversity matters" position to one of compassion, understanding, and acceptance of the salience of race in curriculum, teaching, and learning after the sharing of my racialized narrative. To be clear, the intent of my teaching was not to have my students think in any particular way—or to believe what I believe; the goal was to have the students in the course think about issues of race in education. The narrative was used as a form of the curriculum to assist the students in their quest to better understand themselves in relation to others. Of central concern in the course were questions like (a) who makes curriculum decisions? (b) How are decisions made and on what level? (c) On behalf of whom are curricula decisions made? (d) What roles do economics and social capital play in curriculum decision-making? (e) How are changes made in the curricula? and (f) What influences do teachers have in curriculum development?

At any rate, many students were not initially interested in thinking about race. They appeared skeptical about the relevance and salience of racialized issues and experiences in education. Elsewhere, I have discussed this skepticism, particularly around diversity and race, that I have experienced time and again among students in other courses (Milner, 2005a; Milner & Smithey, 2003). In some ways, I understand (but not condone) the mostly White students' skepticism about the importance and centrality of race in the curriculum, in teaching, and in learning. Kerl (2002) explained that "we cannot necessarily know what is true or even real outside our own understanding of it, our own worldview, our own meanings that are embedded in who we are" (p. 138). Moreover, as Gordon (1990) explained, "critiquing your own assumptions about the world—especially if you believe the world works for you" (p. 88)—is indeed an arduous and complex task.

Thus, the students enrolled in the course did not understand the relevance or the necessity of discussing race in our course because many of them had not experienced racism or inequity in their daily experiences; in other words, the world has worked for them, so many of them wondered why it was necessary to critique and change it in order to better meet the needs of those who are not privileged. To be clear, most of the students in the course probably had experienced some form of inequity (perhaps gender inequity), but they had not thought about or connected that inequity with racialized inequity. It was only after I shared my personal narrative and revealed more of myself that they began to think about their own racialized lives and how that might influence their work. What is it that Black teachers and Black teacher educators often bring into the classroom in their curriculum development that can shed light on how we think about issues of race, curriculum, and teaching?

Black Teachers and Developing the Curriculum

There is a growing and meaningful body of literature that focuses on Black teachers and their teaching (Foster, 1990, 1997; Holmes, 1990; Hudson & Holmes, 1994; Irvine & Irvine, 1983; King, 1993; Milner & Howard, 2004; Milner, 2003; Monroe & Obidah, 2004). This literature is conceptualized in several important ways: it spans the pre-Brown and pre-desegregation era to the present and focuses on P through 12 schools as well as higher education. Understanding the nature of Black teachers and their teaching is especially important in this research because my identity, experiences, and history shaped my curriculum development and teaching. In other words, my ethnic identity helped shape my beliefs and perspectives in my curriculum development.

As asserted, this literature is not limited to public schools; it also highlights Black teachers' experiences in higher education, namely in teacher-education programs (Baszile, 2003; Ladson-Billings, 1996; Milner, & Smithey, 2003). Of central importance in this literature is the question, what does a teacher's racial and cultural background have to do with how he or she develops the curriculum and implement it? As Agee (2004) explained, a Black teacher "brings a desire to construct a unique identity as a teacher . . . she [or he] negotiates and renegotiates that identity" (p. 749) to meet their objectives and to meet the needs of their students. Much of what we know from the literature about Black professors in higher education is that Black teachers often experience resistance and silence from their mostly White students when they focus on what many White students find as uncomfortable issues and perspectives (Ladson-Billings, 1996).

According to hooks (1994), Black female teachers carry with them gendered experiences and perspectives that have been (historically) silenced and marginalized in the discourses about teaching and learning. Although teaching has often been viewed as "women's work," Black women teachers and their worldviews have often been left out of the discussions—even when race was centralized (hooks, 1994). Similarly, in colleges of education, and particularly preservice and inservice programs, the programs are largely tailored to meet the needs of White female teachers (Gay, 2000); hence, Black teachers along with other teachers of color (male and female) are left out of the discussion. What are the needs of Black, Asian, and Latino/a students in teacher-education programs? How are the needs of these students (other than White females) different and being met in teacher-education programs? Indeed, these questions deserve attention as we think about the education of all students in teacher-education programs.

Where curricular materials were concerned—in her study, Agee (2004) explained that "the teacher education texts used in the course made recommendations for using diverse texts or teaching diverse students based on the assumption that preservice teachers are White" (p. 749). Still, Black and other teachers often have distinctive goals, missions, and decision-making and pedagogical styles of teaching that are different from their White counterparts. In her analyses of African American teachers in public schools during segregation, Siddle-Walker (2000) wrote, "Consistently remembered for their high expectations for student success, for their dedication, and for their demanding teaching style, these [Black] teachers appear to have worked with the assumption that their job was to be certain that children learned the material presented" (pp. 265–266).

These teachers worked overtime to help their African American students learn; although these teachers were teaching their students during segregation, they were also preparing their students for a world of integration (Siddle-Walker, 1996). Moreover, as Tillman (2004) suggested, "these teachers saw potential in their Black students, considered them to be intelligent, and were committed to their success" (p. 282). There was something authentic about these Black teachers. They saw their jobs and roles extend far beyond the hallways of the school or their classroom. They had a mission to teach their students because they realized the risks and consequences in store for their students if they did not teach them and if the students did not learn. An undereducated and underprepared Black student, during a time when society did not want nor expect these students to succeed, could lead him or herself to self- and social destruction (drug abuse, prison, or even death).

Among students of any ethnic background, Pang and Gibson (2001) maintained that "Black educators are far more than physical role models, and they bring diverse family histories, value orientations, and experiences to students in the classroom, attributes often not found in textbooks or viewpoints often omitted" (pp. 260–261). Black teachers, similar to all teachers, are texts themselves—they are a form of the curriculum; however, these teachers' pages are inundated with life experiences and histories of racism, sexism, and oppression, along with those of strength, perseverance, and success. Consequently, these teachers' texts are rich and empowering—they have the potential to help students better understand the world (Freire, 1998; Wink, 2000) and some of the complexities of race and racism, for example, in meaningful ways.

African American teachers often felt irrelevant and voiceless in urban and suburban contexts—even when the topic of conversation was multicultural education (Buendia, Gitlin, & Doumbia, 2003; Ladson-Billings, 1996; Milner & Woolfolk Hoy, 2003; Pang & Gibson, 2001). These experiences are unfortunate given the attrition rate of Black teachers in the teaching force, and I am not only referring to the low numbers of Black teachers in public school classrooms but also to the low numbers of Black professors, particularly male professors of teacher education, in higher education. In public schools, Black teachers are leaving the teaching profession and quickly (Hudson & Holmes, 1994). Historically, pre-desegregation (particularly before the 1954 *Brown v. Board of Education* decision), the teaching profession was always viewed as an honorable and popular profession for Blacks (Foster, 1997; Siddle-Walker, 1996).

In their classrooms, Black teachers were able to develop and implement optimal learning opportunities for students—yet in the larger school context, these same teachers were often ridiculed for being too radical or for not being "team players." As evident in my own research (Milner, 2005b), a Black teacher can feel isolated and ostracized because he or she offered a counterstory or counternarrative (Ladson-Billings, 2004; Ladson-Billings & Tate, 1995; Morris, 2004; Parker, 1998; Solorzano, & Yosso, 2001; Tate, 1997) to the pervasive discourses and views of their mostly White colleagues. Because Black teachers' ways of connecting with their students were effective—yet often inconsistent with their non-Black colleagues—these teachers' experiences, inescapably, influenced their teaching in the classroom with students.

Delpit (1995) shared a reaction from a White teacher when talking about the management style and pedagogical approach of a Black teacher: "It's really a shame but she (that Black teacher upstairs) seems to be so authoritarian, so focused on skills and so teacher directed. Those poor kids never seem to be allowed to really express their creativity. (And she even yells at them)" (p. 33).

What the teacher in the passage above seemed to not understand was that the "Black teacher upstairs" may have been quite effective in providing all students access and opportunities to learn. Effective teaching can be quite varied for different teachers and students. The White teacher's criticism of the Black teacher's focus on skills demonstrates how teachers adopt different pedagogical and curricular philosophies and strategies based on their own thinking, beliefs, and teaching situations. Apparently the authoritarian and skills-driven approach by the Black teacher upstairs is an approach that may be effective for that teacher with her students. Black teachers may have a different way of thinking about how best to make learning happen in their classrooms; however, different does not necessarily mean deficit or deficient. In essence, pedagogical and curricula decisions can be racially and culturally mediated. They often depend on the context, and they are not necessarily neutrally constructed. Indeed, the goal in teaching on any level of education should not be to have all teachers think and teach in the same manner or any specific way. Teachers are intellectuals and have to discover and implement their own effective teaching methods in the classroom.

In sum, teachers' in-school and out-of-school experiences—their autobiographies and social realities—influence their curriculum development and teaching with students. In other words, we know that teachers do more than go into a classroom and robotically teach a set of information or materials. Rather, what happens to teachers in their daily lives and experiences (in the supermarket or in a car dealership, for instance) often show up in the curriculum and their teaching. Teachers usually do not divorce themselves from what they believe to be essential for student learning. My racialized experience outside of the classroom emerged in the classroom and enhanced my teaching with my teacher education students.

This Narrative Research

Because narrative research can be described as the study of the stories that people come to experience and live in the world (Clandinin & Connelly, 1996; Connelly & Clandinin, 1990), this research attempts to employ a narrative perspective to understand the influence of my racialized experience in my teaching. Narratives are both the object of investigation and the actual method of investigation (Clandinin & Huber, 2002; Connelly & Clandinin, 1990). Rushton (2004) declared that "lived experiences can be translated into rich narrative stories useful for both teaching and research" (p. 62). My experiences and stories about race deeply shaped my decisions with my students in the college classroom. In a sense, my racialized stories constituted what Carter (1994) conceptualized as "well-remembered events." In my class, my students were experiencing an education that was a "construction and reconstruction of personal and social stories . . . [where] teachers and learners are storytellers and characters in their own and other stories" (Connelly & Clandinin, 1990, p. 2). I attempted to capture the multiple stories around race that I experienced as a way to better understand my own practice (my curriculum decisions) as well as to shed light on the multiple tensions inherent in living and experiencing racism in a multicultural society. Moreover, I attempted to help the students in my course become aware of how their own experiences shape the curriculum.

If we define the curriculum as what students have the opportunity to learn in schools, and if we believe that students learn a combination of information (from what is formally taught as well as what is shared from each other's experiences and from teachers' experiences as told in stories), then we must consider the enormous role race plays in opportunities to learn. Consider, for instance, Eisner's (1994) postulation of several forms of the curriculum: (a) the explicit curriculum concerns student-learning opportunities that are overtly taught and stated or printed in documents, policies, and guidelines, such as in course syllabi; (b) the implicit curriculum is intended or unintended but is not stated or written down; (c) the null curriculum, a third form of curriculum, deals with what students do not have the opportunity to learn. Thus, information and knowledge that are not available for student learning is also a form of the curriculum. What students do not experience becomes messages for the students themselves. For example, if students are not taught to question or to critically examine power structures, the students are learning something—possibly that it may not be essential for them to critique the world in order to improve it. Stories told (or not), then, are a critical part of both the implicit and null curriculum—they influence learning opportunities for students. What is absent or not included is actually present in what students are learning.

Teachers and teacher educators have important authority over what students are exposed to (or not) to in a particular context. To illuminate, Eisner (1994) wrote,

> Teachers inevitably have a range of options that they can exercise in the selection, emphasis, and timing of curricular events. Even when they are expected to follow certain guides or books in which activities and content have already been determined [explicitly], there are still options to be considered and choices to be made by teachers with respect to how those materials will be used and the ways in which what is done in one particular area of study will or will not be related to what is done in other areas of curriculum [implicitly]. These decisions are, of course, decisions bearing on the curriculum; they influence the kind of opportunities for learning and experience that children will have. (p. 126)

What Eisner explained here relates to McCutcheon's (2002) position relative to teachers' decision making. She maintained that teachers transform curriculum policies and materials to such a degree that it is more appropriate to think of them as curriculum developers themselves than mere curriculum implementers.

In addition, I am mindful of the point that "there is little research using [the narrative method] in the multicultural literature" (Connelly, Phillion & He, 2003, p. 365). I was living and working in a space that constantly reminded me just how salient racial tensions still were in our multicultural society, and my teaching had to deal with these tensions. I came to understand the connections that Connelly et al. (2003) made:

"Multiculturalism [and multiracialism] name a way of living and narrative inquiry is a way to think about living" (p. 368). The most useful way for me to describe the nature of my experiences and the essence of my teaching decisions with the students was through the narrative. As Connelly and Clandinin (1990) explained, "People by nature lead storied lives and tell stories of those lives, whereas narrative researchers describe such lives, collect and tell stories of them, and write narrative experience . . . in understanding ourselves and our students educationally, we need an understanding of people with a narrative of life experiences" (pp. 2–3). Accordingly, in a sense, these stories are autobiographical. And this study is one of the self in relation to others.

Whereas a major dimension of narrative research is on studying the self in relation to others (or in relation to participants in a research study), this study focuses mostly on me as I worked to develop a meaningful curriculum for my students in an institution of higher education. The process of conducting this research was similar to Elbaz-Luwisch's (2001) explanation of her own decision making as she engaged in narrative inquiry. She wrote, "Unlike a traditional researcher I did not choose or design the course experience with research in mind. However, like a 'narrative inquirer' I did write and collect field texts: I kept a journal during the course . . . Unlike a traditional researcher I did not bring to bear a particular theoretical framework either before the course began or afterwards when I began analyzing the experience, though I did use various theoretical perspectives to make sense of what happened, to raise questions and to check my ongoing interpretations" (p. 135).

In the next section, I expand on the data collection for this study.

Documenting the Narratives

Similar to Elbaz-Luwisch (2002), the writing or documenting of my stories was a "coming to know" (p. 410). As my stories of experience around race emerged, I was mindful of the impact such experiences had on my curricula and pedagogical decision making in my classes, research, and writing. I began documenting my experiences around race at the university, even before I became fully aware of its influence on my teaching and curriculum decisions. I kept a journal, what I called a "personal reflection notebook," where I wrote about my thoughts and feelings on working in the academy. As my experiences became more illuminating, I started to notice how my race stories "showed up" in my teaching. Much of what I shared in my courses was not planned *per se*. Thus, the data collection method or technique used in the study was a journal or what I called notebook entries.

It is important to note that when I began documenting my experiences, I did not document them for the purpose of this research. I documented my experiences because I often felt overwhelmed as I experienced race and racism in my experiences both inside and outside of the classroom (in the grocery store, at the gas station, in the building at school, in addition to other times and locations). I found the exercise of writing in the notebook therapeutic. The journal was a way for me to vent at times and a way to celebrate in other instances. For instance, I often wrote about successes I had experienced, such as my excitement about publishing research in refereed journals. Thus, the notebook became a valuable tool in this research. Clearly, the story I present in subsequent sections of this chapter is "sustained in my ways of thinking, in my ways of knowing, and in my ways of being" (He, 2002, p. 517).

There was not a particular schedule for documenting my experiences. However, I typically wrote (and continue writing) in my notebook at least once a week. As I document my stories of experience, I do not necessarily write out full pages of text. But rather, I sometimes write key points or themes to help me recreate what has actually happened, when it happened, and perhaps why I think it happened. I always attempt to write about my feelings and thoughts to complement the actual unfolding of events in the notebook. Moreover, I always attempt to recreate the actual experiences in as much depth as possible. In many instances, I try to "demystify" what I experience. I would question, "Am I being too sensitive about this event?" Or, "What did I do to precipitate the occurrence?" Many times, I call friends and colleagues, both inside and outside of higher education, to help me think through my stories. I rationalized my case and then begged for critique: "Am I overreacting to this experience?" "Has anything like this ever happened to you?" Thus, my engagement of reflection in my notebook points succinctly to Clandinin and Huber's (2002) explanation: "We are engaged in living, telling, retelling and reliving our lives within particular social and cultural plotlines" (p. 161). I found myself attempting to retell and relive my experience to make sense of the situation for myself and for others (students, colleagues, and friends).

The Course and Students

The course in which I shared this particular racialized narrative and observed such a positive response from students was called "Curriculum Foundations." The course was required for master's students enrolled in a Curriculum and Instructional Leadership program. Students enrolled in the course were, in large part, inservice teachers who were working on their master's degree part time while teaching full time. In a few instances, students were returning for their masters and did not have any teaching experience. There were eighteen students enrolled: fifteen White women, one Black woman, and two White men.

A Narrative of Experience about Race

Below, I outline an encounter around race that overtly and implicitly shaped my teaching (and consequently my learning) in the classroom with students. After this initial sharing, I presented additional narratives in the class. In this chapter, I chose to focus on one narrative, the initial story, in order to have ample space to analyze and to discuss the meanings behind what occurred. I offer this narrative around race with the desire to help readers understand how race matters (both inside and outside of the classroom) and how influential teachers can be in students' learning opportunities. In essence, the narrative can be a door opener to discuss and think deeply about complex and taboo issues.

Are You the Janitor?

As I walked into the doors of the building to my office, I walked passed the administrative assistant who works in another division. Before I was out of sight, the administrative assistant, a middle-aged White woman whom I had greeted several times in the past, called me back to her work area. "Is that yours," she asked, pointing to the janitorial cart near her workstation. I replied, "Excuse me?" with more than a hint of confusion. "That cart right there—is that yours?" Perplexed, I responded, "No, I am not the janitor." She replied, "Well, whoever it is needs to move it. It is in the way sitting right there." I had greeted—basically said "hello" to the administrative assistant on several occasions in the past. In fact, she always seemed cordial to me as we expressed and exchanged our "hellos" and "good-byes" in the midst of busy days.

Had she not remembered our brief encounters in the past? Was she joking with me? No, this person assumed that I could be the janitor, despite my professional dress. Could I not be a professor? Was I invisible to this administrative assistant in the past? That is, was I encountering what Ellison (1947) referred to as "the invisible man" phenomenon? In other words, was I not noticeable to her? Was I irrelevant and consequently invisible to her? Had she made up in her mind whom I must be? If so, what factors contributed to her decision about whom I must be (my occupation, my role, and my status at the university)?

It is no secret that many of the janitorial staff members in the building are (and have been) African American. It is also no secret that there are very few African American faculty members in the building. I was (and, at present, still am) the only Black professor in the building. What was it that made this person equate me and my presence with that of a janitor? To be clear, I am not degrading or attempting to create a power division between professional and support staff and personnel. I respect the janitorial staff in our building and appreciate their contribution. What I am suggesting is that because I am a Black man, the administrative assistant relied on stereotypical notions and assumptions about who I must be. More than likely, race played a central role in her perceptions of who I was and why I was in the building. Hurt, anger, and frustration were the emotions I felt after this experience.

This Experience and the Curriculum

After assuring the administrative assistant that I was not the janitor and walking upstairs to my office, I called a colleague and friend at another university. I was too baffled to talk extensively about the experience but talked briefly about what had just happened. My colleague assured me that she had experienced similar encounters countless times in her department, at conferences, in hotels, at restaurants, and even in the classroom (as a professor). For instance, this colleague shared experiences when her students acted in disbelief when they discovered that she was indeed their professor for their course. They expected someone else—probably a White professor because as they explained to her they had never had a Black professor in the past.

Because I needed to think deeply about what had just happened, what it meant for me as a teacher/professor and as a Black male in the academy and in education, I needed to engage in deliberation (McCutcheon, 2002) about how I should handle what I had just experienced that had such a profound influence on me (emotionally and intellectually). Because I worked with preservice, inservice, and prospective teachers at the university, I was cognizant of my *duties* and *responsibilities* to make this experience explicit in my curriculum. My colleague and I have written about our role and influence in developing and enacting a curriculum focusing on diversity with preservice teachers. Refer to that article for more on this idea (Milner & Smithey, 2003). Elsewhere, I have written about how African American teachers in public schools often adopted deliberative and purposeful curricula that addressed issues of race and racism in their high-school classes, mainly because these teachers experience race and racism in such meaningful ways outside of the classroom (see Milner, 2003). These African American teachers and I felt a sense of responsibility and duty to centralize issues of race in the classroom because many of our students would experience curricula and instruction that would avoid topics related to race. My experience with the administrative assistant made me feel obligated to share race-based lessons with my students in the college classroom. In other words, when teachers understand and come to know a situation more fully or when they develop deeper understandings of race, they may be more sensitive to those issues in their classrooms with students. Because what teachers know and understand do emerge, either consciously or inadvertently, in the P through 12 classroom, I wanted the students in the college classroom to have the opportunity to broaden their perspectives about race and its importance to their work in education so that ultimately their students might benefit from teachers who were perhaps more racially knowledgeable, sensitive, and aware.

I began taking notes on the implications of my experience for my graduate seminar that would meet that same evening of this incident with the administrative assistant. I deliberately wanted to make my experiences around race an integral part of the curriculum that evening. As I reacquainted myself with the readings for the seminar's meeting, I found ways to introduce and to expand on my narrative. In general, my classes are usually conducted with discussions around a topic. We often share our "stories" and experiences and use them as a theory of experience (hooks, 1994). According to hooks (1994), "As a teacher [educator], I recognize that students . . . enter classrooms within institutions where their voices have been neither heard nor welcomed, whether these students discuss facts-those that any of us might know-or personal experience. My pedagogy has been shaped to respond to this reality. If I do not wish to see these students use the 'authority of experience' as a means of asserting voice, I can circumvent this possible misuse of power by bringing to the classroom pedagogical strategies that affirm their presence, their right to speak, in multiple ways on diverse topics" (p. 84).

In this instance, experience and narrative are theory. In other words, if theories rely on what we know, personal experiences can be viewed as a form of theory—that is, what we come to know. Indeed, my and the preservice teachers' personal experiences were what we knew, came to know (our theories), and what we could draw from and build on in our work. Our racialized narratives can be used as a place to start thinking about issues of race and can be sites to examine and reflect about what we know, how we know it, and why.

Accordingly, in class, we would typically find relevance and connections around teaching and learning to which most of us could relate. The discussion in seminar that evening, however, needed to focus specifically on our differences to locate the connections and similarity among us. That is, my students needed to understand Dillard's (2000) observation that people of color are not White people with pigmented skin. Our experiences, our struggles, and our triumphs are often shaped by who we are as racialized beings. Once we had thought seriously about our differences, we began to think about our similarities. For instance, all people needed to breathe air in order to live (Dillard, 2002). The students and I also could relate to other forms of discrimination that emerged in our daily walks of life. Many of the female students reported gendered experiences after the sharing of my narrative. One student expressed how her adviser during her undergraduate education always encouraged her to take lower-level mathematics courses, despite her high grades in math in high school. The sharing of my narrative brought this student to a space of remembering. She had not thought about this experience in many years but was able to make connections and empathize with me because I shared my story.

The Black student in the class immediately understood my narrative and agreed with me as I discussed the importance of understanding how racism emerges in education (curriculum, teaching, and learning). In fact, she seemed to understand from the outset of the course. She knew about racism and understood it because of her lived experiences. Moreover, all of my students needed to see how important it could be for teachers, on any level of education, to be conscious of their narratives because these narratives have the potential to emerge in the classroom to shed light on situations and to serve as opportunities to discuss, research, and learn—just as my narrative had shown up in my teaching.

Although many of the students began to remember discriminatory experiences, after I shared my narrative, still, it was no easy feat for me to address and to have them discuss and reflect about issues of race and racism with my students. However, the narrative can tear down tension in situations and bring about understanding, discussion, and empathy (Blake, Stout, & Willet, 2004). There was what appeared to be an "opening up" of sorts as the students in my class heard the narrative. Prior to my sharing of the story, many of the students appeared resistant, reluctant—even skeptical—about the seriousness, relevance, and applicability of race and racism in curriculum, teaching, and learning. Sharing my narrative, however, seemed to open and to promote a safe space where students could "let down their guard" and open themselves up to new possibilities and new beginnings where race and racism were concerned, as they thought about their own experiences and discrimination.

My curriculum and pedagogy in seminar that evening was infiltrated with that racialized experience. The preestablished theme on the syllabus for the evening was "teachers as curriculum developers." I asked the students to think about their own beliefs about race and culture and for them to connect these beliefs to their own curriculum development in public schools. I asked the students to relate the readings for that evening to race and culture. In short, I asked the students to think about the previously scheduled focus of the class through cultural and racial lenses. I realized in that class session, perhaps more so than ever in the past, that teaching and learning truly are about personal encounters. We focused on several important areas of the curriculum and instruction in thinking about race. For instance, we discussed what students had the opportunity to learn in classrooms (the curriculum) and who decides what knowledge is actually covered. We also considered how teachers can take the stated, explicit, predetermined, or established curriculum, such as the syllabus outline for the seminar that evening, and redirect the curriculum to meet perhaps more pressing needs. We discussed the enormous amount of power that teachers and teacher educators have. We discussed some of the tensions in the use of this power and how teachers can empower their students on various levels of education and schooling. In a sense, I attempted to model what could (and perhaps should) happen in P through 12 classrooms.

Teaching is often filled with what we as teachers experience outside of the classroom, and we often use examples—narratives—from those experiences in the classroom with students as forms of the curriculum. In essence, I attempted to model my own decision making and shared my sense of authority so that the teacher-education students could recognize their own influences and authority on the curriculum with their own students. I modeled my curriculum development by actually explaining my thinking to students in the decision-making process. For instance, I shared with them the processes I went through after the experience with the administrative assistant. I explained that I called and shared my story with a colleague/friend, and I discussed how I redirected my lesson plans for the evening's seminar. I explained that they, too, had the potential to alter their work with students using their personal narratives, and I encouraged them to invite their students in public schools to share their narratives as well with special focus on race where appropriate. We also discussed some of the problems and tensions with this authority. We determined that the goal should be to empower, not to dictate, what and how things should be.

After sharing my narrative with the students, many of them offered empathetic comments: "Wow—how insensitive and inconsiderate." "Why would she think you were the janitor?" "What was she [the administrative assistant] thinking?" Students in the course also mentioned how they would have felt having experienced what I shared. These emotions varied: "I would have been so hurt." "I would have been very upset—quite angry." This is not to suggest that all students were on board with the relevance and salience of race and racism in society and thus in education. Some "got it" right off and seemed much more open to the centrality of race and racism in society and in our schools. To be clear, the point of my sharing of the story was for the students (fundamentally) to consider whether or to what extent racism and race still mattered (West, 1993) in our society and thus in education. Because one of the main objectives of the course was to make connections between society and schools (that is, to understand how what happens in society influences what happens in the classroom curriculum), the students were able to understand how some of their own students could experience life in society and how those experiences could shape their beliefs and thinking inside of the classroom. A small number continued to be resistant, did not want to think about race, and seemed offended that I would even share the narrative: "We have come a long way in society. I am thankful that things are not like they used to be." Such statements made me remember that if one is in power and lives from a privileged position (McIntosh, 1990), then it is increasingly difficult for that individual to understand others' marginalization and mistreatment. Delpit's (1995) work echoed in my mind as I drove home from class that evening. Activities and goals of courses like this—where students are not asked to think in any certain way about a topic but are asked to think about an issue regardless of what they conclude—should focus on progress that the students are making through the process of engaging in deep thinking and introspection and not necessarily on what they conclude.

Delpit described five aspects of power:

> (a) issues of power are enacted in classrooms; (b) there are codes or rules for participating in power; that is, there is a "culture of power;" (c) the rules of the culture of power are a reflection of the rules of the culture of those who have power; (d) if you are not already a participant in the culture of power, being told explicitly the rules of that culture makes acquiring power easier; and (e) *those with power are frequently least aware of—or least willing to acknowledge—its existence. Those with less power are often most aware of its existence.* (p. 24, emphasis added)

Clearly, although I had some power as "professor" in the course, my students in the class had power (by virtue of their skin tone) that could easily override my power. A simple visit to the dean's office could make the dean skeptical about me or possibly even think I was a racist because I made some of the students feel "intellectually or socially uncomfortable."

Moreover, they had the power to drop the course and to give me low teaching evaluations. The students also had the power to leave that classroom context and go into society—even the hallway—and regain their power and privilege. They had power because they were White. The students had the power and the privilege to "tune me out" or to counter my every position.

Fortunately, most students seemed to open their minds and hearts (Banks, 1998) to engage in thinking, discussion, and understanding about race and racism in education. As Conle (2003) explained, "During this act of telling, [individuals] are likely to remember or discover incidents and details not held in mind at the outset of the telling. This remembering may in fact even change their story or prompt a new understanding of it" (p. 7). The students in my class began to share their own stories and stories of their friends that pointed succinctly to race, racism, and (in)equity. Most importantly, my telling of this narrative brought to my consciousness other race-central stories that I needed to share, mainly because I seemed to have broken down some barriers simply through the telling (my sharing) of that narrative, and students became willing to engage in the intellectual work necessary to learn more and to become more knowledgeable. It appeared that students wanted to hear additional stories, and they shared their own narratives. In short, I began to see "the light in their eyes" (Nieto, 1999). I was able to share other stories that also had the potential to shed even more light on the complex nature of curricula, teaching, learning, and race. The students were thinking, which was a first step to more deeply understanding for the sake of their practice with students in P through 12 classrooms.

Importance of Narrative and Self-Study

The sharing of my experiences seemed to resonate with the students intellectually (their thinking) and their hearts (their affect, care, and concern; Banks, 1998; Elbaz-Luwisch, 2001; Milner, 2005a). There was something important that occurred when the students saw me as a *real* person who experienced life—the good as well as the bad. In many ways, it was the experiences that we shared (Dillard, 2002) that was the conduit for thinking, discussing, and understanding race and racism. My students in the class had all experienced hurtful experiences—clearly not the same as my racialized hurtful experience—but they knew what it felt like to hurt. It was this connection that the narrative provided that had such an impact on helping the students want to engage in thinking and discussions about race. As Conle (2003) declared, the narrative "encompasses not only what is explicitly learned but also what is learned practically, at a more tacit level, touching not only the intellect but the moral, practical, imaginative realm" (p. 3). The sharing of my narrative seemed to push my students to engage in introspection that caused many of them to rethink what they thought they knew about race, teaching, learning, and the curriculum. For instance, many of the students reported that they did not believe racism still existed in the United States. Others reported that they had never had teachers or friends who were of a different race. Some reported that I was their very first Black teacher or professor. Many of the students genuinely seemed to be either unaware or misinformed about the pervasiveness and prevalence of racism in society and consequently in schools. Their experiences simply did not allow them opportunities to think about such issues. After the sharing of the race narrative and after some serious discussion, many of the students started to recall incidents that they had not considered as racist or race central before; their reflection allowed them to come to terms with some of these issues. For instance, one student (who had initially expressed her opinion that racism had ended) later—after the sharing of my narrative—explained that her grandmother often used racist remarks that made her feel uncomfortable. Overwhelmingly, when asked if their parents or grandparents would agree with their dating or marrying someone outside of their race, most of the students reported that their family would have problems with this. Still, some of these same students found it difficult to believe that race and racism were so important in education. They somehow believed that racism disappeared in the classroom even though they saw it in their own homes (when their parents expressed their positions on cross-ethnic dating).

Indeed, the sharing of race-related narratives can prove meaningful and productive in helping students and teachers understand, think about, and change their thinking about such issues. Once students (and teacher educators) *know* better, they are more likely to *do* better. However, McVee (2004) also explained that "narratives can be educative, but the work of such learning is difficult and requires thoughtful curriculum, teaching, and time" (p. 897). Thus, the real test of the effectiveness of the narrative comes when we are able to document improvement in teachers' practices with students, particularly students of color. Teacher educators need to engage in more self-study and to follow their students (preservice and inservice teachers) into the classroom to capture their curricula and pedagogical decisions with students in P through 12 schools and document the effectiveness of our (teacher educators') teaching in the teacher-education classroom. Hopefully, we are providing spaces for our students to engage in thinking and discussion that helps them make more informed curricular and instructional decisions.

Clearly, the self-study research approach is critical to improving practices in teacher education. Hamilton and Pinnegar (2000) explained that "by engaging in self-study, every teacher educator would become a scholar of teacher education, of whatever phase or aspect of teacher education might be intriguing or revealing of a living contradiction in his or her beliefs about teacher education" (p. 239). Toll, Nierstheimer, Lenski, and Kolloff's (2004) research explored their own stories about teaching to problematize how they responded to their preservice elementary-education teachers. In essence, Toll, et al. (2004) explained that they "had the urge to cleanse students of the beliefs and practices they [brought] back from [their] field [experiences]" (p. 164). For instance, the authors shared that "even though we consider ourselves to be constructivists, we were disturbed when our students constructed meaning with which we disagreed" (p. 165). The teacher educators used stories, both telling and listening, to help them think through the conflicts that they encountered with the elementary teachers. Similarly, I found myself wanting my students to broaden their knowledge about race and to change their beliefs and actions where racism, inequity, and oppression were concerned. However, I did not want to misuse my authority over the students and force them to think in any *certain* way about the matters I found most salient in the class. Rather, similar to Toll et al. (2004), I wanted my students to develop their own meanings about these issues and to change their racist views and perspectives. At the same time, I also had a set of objectives and goals (both implicit and explicit) that I hoped the students would master. Thus, without self-study, it is difficult for teacher educators to learn from their work as they work to improve it and to make a difference for students in P through 12 schools.

Attempting to balance my own perspectives, beliefs, and objectives with my students' varying perspectives was not an easy feat. As Elbaz-Luwisch (2002) explained, "Teaching stories are in part personal stories shaped by the knowledge, values, feelings, and purposes of the individual teacher . . . teachers are called upon to mediate constantly between their personal understandings, values, and commitments, and the external requirements of teaching elaborated by policymakers, administrators, parents, and members of the public, all of whom stake claims on the contested social practice of teaching" (p. 405).

Of particular importance for teacher educators is how they balance their beliefs, values, and positions with the needs of students in their higher-education courses *as well as* the needs of students in P through 12 classrooms. Merryfield (2000) questioned whether teacher educators possess what is necessary to teach for equity when many have not "had even the minimal kind of experiences with diverse cultures or the basic understandings of inequities" (p. 430). Self-study is essential to understanding our (teacher educators') positions about teaching, learning, and education—and especially about racism.

Conle (2003) maintained that "the act of narrating therefore is tremendously important. We may understand our own story differently and create different narrative statements about a particular story in different situations" (p. 6). The self-study line of research is extremely important as is the narrative approach to studying our work. More studies need to combine these two potentially ground-breaking approaches in order to improve what we know (theoretically) and how we know it (methodologically) in teacher education. The results of such studies can potentially provide profound insights.

At the end of Alvine's (2001) courses, "students unanimously point to the autobiographical narrative as the most important task of the semester and encourage [her] to be sure to keep that assignment in [her] syllabus" (p. 10). My students also expressed how meaningful the sharing of narratives was for them as well in our course. Thus, I agree with Klein (2004) that attempting to raise awareness is one of the most challenging tasks I face as a teacher educator; employing the narrative seemed to make a difference in raising awareness about race and racism in education, and the students even found the value in using the narrative to shape their own experiences.

I agree with Elbaz-Luwisch (2001) that "storytelling can be a way of admitting the other into one's world and thus of neutralizing the otherness and strangeness" (p. 134). I was granting my students entry into my experiences—my life—that shaped their own openness and hopefully their own teaching. Moreover, I offered a counterstory or counternarrative (Ladson-Billings, 2004; Ladson-Billings & Tate, 1995; Morris, 2004; Parker, 1998; Solorzano & Yosso, 2001; Tate, 1997) to the pervasive views of my mostly White students who live, act, and experience a world quite different from and inconsistent with my own. And, at the same time, I offered hope to the Black student in the class who shared with me in a private meeting that she often felt ostracized by her White classmates when she offered her counterstories where race and diversity were concerned.

Elbaz-Luwisch (2001) argued that "telling our stories is indeed a matter of survival: only by telling and listening, storying and restorying can we begin the process of constructing a common world" (p. 145). I agree with Conle (2003) when she explains, "It is the narrative repertoire of our imagination that helps us distinguish the world we live in from the world we *want* to live in" (p. 4, emphasis added).

References

Agee, J. (2004). Negotiating a teaching identity: An African American teacher's struggle to teach in test-driven contexts. *Teachers College Record, 106*(4), 747–774.

Alvine, L. (2001). Shaping the teaching self through autobiographical narrative. *The High School Journal, 84*(3), 5–12.

Banks, J. A. (1998). Curriculum transformation. In J. A. Banks (Ed.), *An introduction to multicultural education* (2nd ed., pp. 21–34). Boston, MA: Allyn and Bacon.

Baszile, D. T. (2003). Who does she think she is? Growing up nationalist and ending up teaching race in White space. *Journal of Curriculum Theorizing, 19*(3), 25–37.

Bell, D. (1992). *Faces at the bottom of the well: The permanence of racism*. New York: Basic Books.

Blake, M. E., Stout, L., & Willet, C. (2004). Personal narratives: A tale of three stories. *New England Reading Association Journal, 40*(10), 54–61.

Buendia, E., Gitlin, A., & Doumbia, F. (2003). Working the pedagogical borderlands: An African critical pedagogue teaching within an ESL context. *Curriculum Inquiry, 33*(3), 291–320.

Carter, K. (1994). Preservice teachers' well-remembered events and the acquisition of event structured knowledge. *Journal of Curriculum Studies, 26*(3): 235–252.

Clandinin, J. D., & Connelly, M. F. (1996). Teachers' professional knowledge landscapes: teacher stories-stories of teachers-school stories-stories of schools. *Educational Researcher, 25*, 24–30.

Clandinin, J. D., & Huber, J. (2002). Narrative inquiry: Toward understanding life's artistry. *Curriculum Inquiry, 32*(2), 161–169.

Cochran-Smith, M. (2000). Blind vision: Unlearning racism in teacher education. *Harvard Educational Review, 70*(2), 157–191.

Conle, C. (2003). An anatomy of narrative curricula. *Educational Researcher, 32*(3), 3–15.

Connelly, M. F., & Clandinin, J. D. (1990). Stories of experience and narrative inquiry. *Educational Researcher, 19*, 2–14.

Connelly, M. F., & Clandinin, J. D. (2000). Narrative understandings of teacher knowledge. *Journal of Curriculum and Supervision, 15*(4), 315–331.

Connelly, M. F., Phillion, J., & He, M. (2003). An exploration of narrative inquiry into multiculturalism in education: Reflecting on two decades of research in an inner-city Canadian community school. *Curriculum Inquiry, 33*(4), 363–384.

Delpit, L. (1995). *Other people's children: Cultural conflict in the classroom*. New York: The New Press.

Dillard, C. B. (2002). Walking ourselves back home: The education of teachers with/in the world. *Journal of Teacher Education, 53*(5), 383–392.

Dillard, C. B. (2000, April). Cultural consideration in paradigm proliferation. Paper presented at the annual meeting of the American Educational Research Association, New Orleans, LA.

Dinkelman, T. (2000). An inquiry into the development of critical reflection in secondary student teachers. *Teaching and Teacher Education, 16*, 195–222.

Dome, N., Prado-Olmos, P., Ulanoff, S. H., Garcia Ramos, R. G., Vega-Castaneda, L., Quiocho, A. M. L. (2005). "I don't like not knowing how the world works": Examining preservice teachers' narrative reflections. *Teacher Education Quarterly, 32*(2), 63–83.

Eisner, E. W. (1994). *The educational imagination: On the design and evaluation of school programs*. New York: MacMillan College Publishing Company.

Elbaz-Luwisch, F. (2001). Understanding what goes on in the heart and the mind: Learning about diversity and co-existence through storytelling. *Teaching and Teacher Education, 17*, 133–146.

Elbaz-Luwisch, F. (2002). Writing as inquiry: Storying the teaching self in writing workshops. *Curriculum Inquiry, 32*(4), 403–428.

Ellison, R. (1947). *Invisible man*. New York: Vintage International.

Foster, M. (1990). The politics of race: Through the eyes of African-American teachers. *Journal of Education, 172*, 123–141.

Foster, M. (1997). *Black teachers on teaching*. New York: The New Press.

Freire, P. (1998). *Pedagogy of the oppressed*. New York: Continuum.

Gay, G. (2000). *Culturally responsive teaching: Theory, research, & practice*. New York: Teachers College Press.

Gordon, B. M. (1990). The necessity of African-American epistemology for educational theory and practice. *Journal of Education, 172*(3), 88–106.

Hamilton, M. L., & Pinnegar, S. (2000). On the threshold of a new century: Trustworthiness, integrity, and self-study in teacher education. *Journal of Teacher Education, 51*(3), 234–240.

Hatton, N., & Smith, D. (1995). Reflection in teacher education: Towards definition and implementation. *Teaching and Teacher Education, 11*(1), 33–49.

He, M. (2002). A narrative inquiry of cross-cultural lives: Lives in the North American academy. *Journal of Curriculum Studies, 34*(5), 513–533.

Holmes, B. J. (1990). New strategies are needed to produce minority teachers (Guest Commentary). In A. Dorman (Ed.), *Recruiting and retaining minority teachers*. Oak Brook, IL: North Central Regional Educational Laboratory.

Hopper, T., & Sanford, K. (2004). Representing multiple perspectives of self-as-teacher: School integrated teacher education and self-study. *Teacher Education Quarterly, 31*(2), 57–74.

hooks, b. (1994). *Teaching to transgress: Education as the practice of freedom*. New York: Routledge.

Hudson, M. J., & Holmes, B. J. (1994). Missing teachers, impaired communities: The unanticipated consequences of Brown v. Board of Education on the African American teaching force at the precollegiate level. *The Journal of Negro Education, 63*, 388–393.

Irvine, R. W., & Irvine, J. J. (1983). The impact of the desegregation process on the education of Black students: Key variables. *The Journal of Negro Education, 52*, 410–422.

Kerl, S. B. (2002). Using narrative approaches to teach multicultural counseling. *Journal of Multicultural Counseling Development, 30*(2), 135–143.

Kienholz, K. (2002). Let me tell you a story: Teacher lore and pre-service teachers. *Action in Teacher Education, 24*(3), 37–42.

King, S. (1993). The limited presence of African-American teachers. *Review of Educational Research, 63*(2), 115–149.

Klein, A. M. (2004). Narrative history in the college classroom: Re-contextualizing multicultural issues through close re-reading of juvenile literature. *Multicultural Education, 12*(2), 51–54.

Ladson-Billings, G. (1996). Silences as weapons: Challenges of a Black professor teaching White students. *Theory into Practice, 35*, 79–85.

Ladson-Billings, G. (1998). Just what is critical race theory and what's it doing in a nice field like education? *Qualitative Studies in Education, 11*(1), 7–24.

Ladson-Billings, G. (2004). New directions in multicultural education: Complexities, boundaries, and critical race theory. In J. A. Banks & C. A. M. Banks (Eds.), *Handbook of research on multicultural education* (2nd ed., pp. 50–65). San Francisco, CA: Jossey-Bass.

Ladson-Billings, G., & Tate, B. (1995). Toward a critical race theory of education. *Teachers College Record, 97*(1), 47–67.

McCutcheon, G. (2002). *Developing the curriculum.* Troy, NY: Educators' Press International.

McIntosh, P. (1990). White privilege: Unpacking the invisible knapsack. *Independent School, 90*(49), 31–36.

McVee, M. B. (2004). Narrative and the exploration of culture in teachers' discussions of literacy, identity, self, and other. *Teaching and Teacher Education, 20*, 881–899.

Merryfield, M. M. (2000). Why aren't teachers being prepared to teach for diversity, equity, and global interconnectedness? A study of lived experiences in the making of multicultural and global educators? *Teaching and Teacher Education, 16*, 429–443.

Milner, H. R. (2003). A case study of an African American English teacher's cultural comprehensive knowledge and (self) reflective planning. *Journal of Curriculum and Supervision, 18*(2), 175–196.

Milner, H. R. (2005a). Stability and change in prospective teachers' beliefs and decisions about diversity and learning to teach. *Teaching and Teacher Education, 21*(7), 767–786.

Milner, H. R. (2005b) Developing a multicultural curriculum in a predominantly White teaching context: Lessons from an African American teacher in a suburban English classroom. *Curriculum Inquiry, 35*(4), 391–428.

Milner, H. R., & Howard, T. C. (2004). Black teachers, Black students, Black communities and *Brown*: Perspectives and insights from experts. *Journal of Negro Education, 73*(3), 285–297.

Milner, H. R., & Smithey, M. (2003). How teacher educators created a course curriculum to challenge and enhance preservice teachers' thinking and experience with diversity. *Teaching Education, 14*(3), 293–305.

Milner, H. R., & Woolfolk Hoy, A. (2003). A case study of an African American teacher's self-efficacy, stereo-type threat, and persistence. *Teaching and Teacher Education, 19*, 263–276.

Monroe, C. R., & Obidah, J. E. (2004). The influence of cultural synchronization on a teacher's perceptions of disruption: A case study of an African-American middle-school classroom. *Journal of Teacher Education, 55*(3), 256–268.

Morris, J. E. (2004). Can anything good come from Nazareth? Race, class, and African American schooling and community in the urban south and Midwest. *American Educational Research Journal, 41*(1), 69–112.

Nieto, S. (1999). *The light in their eyes: Creating multicultural learning communities.* New York: Teachers College Press.

Pang, V. O., & Gibson, R. (2001). Concepts of democracy and citizenship: Views of African American teachers. *The Social Studies, 92*(6), 260–266.

Parker, L. (1998). Race is . . . race ain't": An exploration of the utility of critical race theory in qualitative research in education. *Qualitative Studies in Education, 11*(1), 45–55.

Phillion, J., & Connelly, F. M. (2004). Narrative, diversity, and teacher education. *Teaching and Teacher Education, 20*, 457–471.

Pinar, W. F. (2001). *The gender of racial politics and violence in America: Lynching, prison rape, and the crisis of masculinity.* New York: Peter Lang.

Pinnegar, S. (1996, Fall). Sharing stories: A teacher educator accounts for narrative in her teaching. *Action in Teacher Education, 18*, 13–22.

Rearick, M. L., & Feldman, A. (1999). Orientations, purposes and reflection: A framework for understanding action research. *Teaching and Teacher Education, 15*, 333–349.

Rushton, S. P. (2004). Using narrative inquiry to understand a student-teacher's practical knowledge while teaching in an inner-city school. *The Urban Review, 36*(1), 61–79.

Schwarz, G. (2001). Using teacher narrative research in teacher development. *The Teacher Educator, 37*(1), 37–48.

Siddle-Walker, V. (1996). *Their highest potential: An African American school community in the segregated South.* Chapel Hill, NC: University of North Carolina Press.

Siddle-Walker, V. (2000). Valued segregated schools for African American children in the South, 1935–1969: A review of common themes and characteristics. *Review of Educational Research, 70*(3), 253–285.

Solorzano, D. G., & Yosso, T. J. (2001). From racial stereotyping and deficit discourse toward a critical race theory in teacher education. *Multicultural Education, 9*(1), 2–8.

Tate, W. F. (1997). Critical race theory and education: History, theory and implications. In M. Apple (Ed.), *Review of research in education* (pp. 195–247). Washington, DC: American Educational Research Association.

Tillman, L. C. (2004). (Un)Intended consequences? The impact of Brown v. Board of Education decision on the employment status of Black educators. *Education and Urban Society, 36*(3), 280–303.

Toll, C. A., Nierstheimer, S. L., Lenski, S. D., & Kolloff, P. B. (2004). Washing our students clean: Internal conflicts in response to preservice teachers' beliefs and practices. *Journal of Teacher Education, 55*(2), 164–176.

Valdez, A., Young, B., & Hicks, S. J. (2000). Preservice teachers' stories: Content and context. *Teacher Education Quarterly, 27*(1), 39–58.

West, C. (1993). *Race matters.* Boston: Beacon Press.

Wink, J. (2000). *Critical pedagogy: Notes from the real world.* (2nd ed.). New York: Longman.

9

Messages to Teacher Educators from the Margins

Teachers of Color on Equity in Diverse Classrooms

Beverly Cross, University of Memphis

At a recent annual meeting of the American Educational Research Association (AERA), one of the nation's most prominent African American scholars conceded to me that trumpeting the battle cry for racial and ethnic diversity in the teaching force is at a near loss. She stated, "In some cases the opposite is true—the teaching force is becoming less diverse year after year. And in some cases, the increase in teacher diversity is only slight." She concluded, somewhat somberly, that teacher educators should redirect much of their energy to take to scale some of the promising work under way to explicitly and directly better prepare a White national teaching force to teach in diverse classrooms and schools. She stated, "We should be honest that most teacher educators are preparing White teachers for diverse classrooms and schools, and we should be exact in doing so rather than imagining we are creating a diverse teaching force." When she heard me gasp for air as if I had the wind knocked out of me, she kindly and sympathetically stated, "We do not have to totally abandon the idea because the children in these classrooms depend on us to give them an experience with teachers who look like themselves." This is a scholar who has dedicated so much of her work to the struggle to diversify the teaching force, yet her deduction seemed somewhat resolute and distressing. Although she stated we should not abandon the work, her words delivered a jolting blow for American education and for what Banks (2006) referred to as the demographic imperative: "The significant changes in the racial, ethnic, and language groups that make up the nation's population" (p. 155). He articulated the context that establishes this imperative: "The U.S. Census projects that people of color will make up 50 percent of the nation's population by 2050. Students of color make up 40 percent of the Nation's school population in 2004 and are projected to make up about 48% of the nation's school age by 2020 (Cites Pallas, Natriellllo, & McDill, 1989). The U.S. Census indicated that 18.4 percent of school-aged youth spoke languages other than English at home" (p. 155).

Of course, there is much written about the exigency of this demographic imperative and the recognition that it simultaneously creates a school and classroom mismatch between a White teaching force and a diverse student population. Scholars define this mismatch in terms of racial, ethnic, and cultural background; lived experiences; and societal realities (Au & Jordan, 1981; Jordan, 1985; Vogt, Jordan, & Tharp, 1987). The mismatch is important because it brings to question the role of education as the great equalizer. Many in the general public view education as the equalizer (education as leveling the playing field), while some scholars view this equalizer perspective as a myth. The mismatch can be considered one contributing factor in maintaining the education myth and inequity. For example, the mismatch can be viewed as either the cause or the effect of seemingly intractable academic-achievement gaps, the cause or effect of low student motivation, the cause or effect of high dropout rates, the cause or effect of swelling educational nihilism, the cause or effect of high teacher attrition, and the cause or effect of low graduation rates. The intent of this chapter is not to decide these cause or effect dilemmas but rather to focus specifically on what the mismatch means in relationship to either the stagnant plateau or decline in diversity in the teaching force at the same time that diversity in the student population increases precipitously.

It has been several years since that conversation at AERA and several years since I began to consider the conversation in relation to Bank's demographic imperative. Yet it has remained in my consciousness, resulting in personal and professional restlessness leading to a series of questions and scholarly ponderings. At other times, it has existed in my subconscious resulting in a nagging diversion while I pursue other questions and research. But it has never been forgotten. Finally, it led me to question, research, and isolate what was so troublesome to me about that very brief conversation. The question of *what our nation and our educational system gives up if we never achieve diversity—and particularly the diversity that Banks suggests above—in the teaching force* became recursive in my scholarly thoughts. Is that symbolically giving up on the diversity of the nation itself and any chance of eliminating a major paradox in our society? Is it yet another example of what Myrdal (1944) called the "American dilemma" or the gap between U.S. ideals and practices? What does it mean that the teaching force will continue to be overwhelming White while the students will increasingly come from racial, ethnic, and linguistic minority groups? What does the loss of diverse voices, practices, and perspectives in the teaching force mean and how can we continue to value, honor, and maintain these in the face of their physical absence and thus their marginalization from the major work left to be done to minimize education's role in maintaining the American dilemma.

One way to assure that these voices continue to have a presence and influence in education is to add to the rich scholarly work that has been done to insert the voices of teachers of color into educational research, debate, dialogue, and decision making. Therefore, I set out on a scholarly endeavor to explore what teachers of color would say to teacher educators (who are charged with producing the nation's teachers) in the context of the demographic imperative, which is simultaneously an educational-equity imperative. That is, what would this group of teachers ask teacher educators to incorporate into teacher-preparation programs considering we will shortly face an overwhelmingly White teacher-education force preparing an overwhelmingly White teaching force to teach increasingly racially, ethnically, and linguistically diverse student populations? In particular, what would teachers of color leave as their legacy to the preparation of White teachers for diverse classrooms and schools? It is not exactly a dramatization to state that their request is nearly posthumous, but it might come close to being so.

What is reported here is the result of a larger year-long, qualitative study designed to explore the relationship between and among the racial identities of three teachers of color, their curricula, and their teaching practices. Within that study, each teacher identified numerous messages to teacher educators about preparing future teachers based on their own experiences as racial, ethnic, and linguistic minority teachers who teach in diverse classrooms and schools. The study included an African American, a Latina, and a Native American teacher. The intent here is not to generalize these voices or to particularize them. Rather the aim is to elucidate the subjective and multiple experiences and understandings of the three teachers as a means to allow familiarity, comfort, and real-world conditions to inform theory (Patton, 2002). In what follows, I attempt to reinsert the voices of a critical group of the education community that is slowly becoming invisible as their presence in education ebbs to dangerous lows. But before I report the findings, I will provide a brief analysis of the demographic imperative and its importance for the educational-equity imperative.

The Exigency of the Reality

For several years now, it has been feared that the decline in the numbers of teachers of color in the United States would likely reach a statistically immeasurable level, while it has been certain that the student population would change in the opposite direction. In 2005, U.S. classrooms were populated by nearly 42 percent students of color while the teaching force was populated by 83.3 percent White teachers (NCES, 2007). It is already the case that this context is even more severe in large urban school districts populated by large numbers of racial, ethnic, and linguistic minority students. The racial, ethnic, and linguistic gap or mismatch is evident at one level through education statistics like those just presented. The meaning of this gap goes well beyond what the mere statistics indicate and was explored in a recent American Association of Colleges of Teacher Education (AACTE) Wingspread Conference (AACTE, 2004) where the participating national educational leaders concluded that "the nation's culturally and linguistically diverse classrooms face a critical shortage of high-quality teachers. Rapid demographic changes in the P–12 student population as well as the widening achievement gap between White students and students of other racial/ethnic and linguistic backgrounds make the need for a multicultural teaching force even more urgent than in the past" (p. 5).

It is important to understand that this is not a mere call for diversity in the teaching force because it is a good thing to do. Rather the evidence paints a picture about opportunity to learn for students of color in this climate. The AACTE Wingspread leaders (2004) go on to explain,

> An increasing body of evidence shows that teachers with similar racial, ethnic, and linguistic backgrounds to their students have a positive impact on student academic achievement. A recent large-scale study (Clewell, Puma, & McCay, 2001) shows that for Hispanic and Black students (particularly for Hispanic) having a teacher of the same race or ethnicity results in increased test score gains in reading and mathematics. Other studies show that at urban schools with larger numbers of minority teachers, minority students are more likely to graduate and pursue a college education (Hess & Leal, 1997). The mere presence of a teacher from the same racial/ethnic group improves students' motivation to succeed (Meier & Stewart, 1991). (p. 7)

Considering the critical role of teachers of color on the academic achievement of students of color, it is critical that the education community learn more about what these teachers do and subsequently incorporate those practices into teacher-preparation programs. Siddle-Walker (2001) and Foster (1994) point to the dearth of a comprehensive body of knowledge on African American teaching. Such a knowledge base is also incomplete for other teachers of color. Without any substantial presence of teachers of color, there is sure to be an effect on the academic achievement for students already believed to lag behind their White counterparts. It appears that students of color are being deprived of yet another academic support system—teachers of their own ethnic, racial and linguistic backgrounds—added to the lack of support through experienced teachers, cutting-edge technologies, advanced academic courses, and high expectations.

In addition to academic achievement, the lack of diversity in the teaching force will also mean that students will confront the devaluing of the richness of cultural knowledge (Gordon, 1995) and how that knowledge can serve as a foundation for academic achievement (Valenzuela, 1999). For example, research on the teaching crisis in California (with 60 percent of students being students of color and nearly four out of five teachers being White) concluded that "though being academically proficient in teaching does not depend on one's race, the ability to understand and relate to students often has everything to do with race" (Keleher, Piana, & Fata, 1999, p. 4). The study further identifies four "key reasons that students of color stay in school longer and achieve more when they have teachers who share their racial and cultural experience." These reasons are (a) the role-model effect, (b) the power of expectations, (c) cultural relevance, and (d) teacher retention. Such connections are well established by scholars as essential to educational achievement and equity. This context has been examined by scholars through the lenses of culturally relevant and responsive pedagogy (Gay, 2000; Irvine & Armento, 2001; Ladson-Billings, 1994), linguistic diversity (Valenzuela, 1999), and multicultural education (Banks, 2006). These scholars demonstrate clearly that a diverse teaching force is essential to equity in education for our nation's diverse student population.

This study adds the voices of three practicing racial-minority teachers who teach in large, urban schools. The teachers represent three different racial groups, represent both elementary and high-school teachers, and are all female teachers. These teachers were selected because they are racial-minority teachers who are recognized in their school district for the relationships they have with their students and for the high achievement of their students. They also represent the four connections identified above (the role model effect, the power of expectations, cultural relevance, and teacher retention). What is presented in the next section is a short letter from each of the three teachers presenting their recommendations to teacher educators. They know from their own observations that their group likely may become extinct in the future, since there is little evidence that they will be replaced by other teachers of color when they retire. But their voices and perspectives on preparing future teachers do not have to become extinct. They can be heard through these messages and through the work of others to assure their voices and ideas are not lost. The messages have been compiled from individual interviews conducted over a year with each teacher. Although their ideas and recommendations for teacher educators are numerous, I have selected five key ideas to include here from each teacher. Each message represents the words of the teachers only—they are direct excerpts from the interviews. I chose to present this research in letter format in honor of *A Talk to Teachers* by James Baldwin (1963/1985). He referred to the 1960s as a dangerous time in education and called on teachers to "go for broke" (p. 1). These teachers, too, recognize the exigency of the context in which they currently teach as well as the one we will face when others like them may have an immeasurable presence in education. Following these individual messages, I will explore the implications that surfaced among the teachers.

Messages to Teacher Educators from the Margins

The following messages are written by minority teachers to teacher educators. They are based on the experiences of three teachers of color in large, urban schools. They are intended to communicate critical issues in teacher preparation from the perspectives of these teachers. It is clear that they are very dedicated and thoughtful about teaching. What they state may not be entirely unique or new, but it is powerful in its derivation within the classroom and from voices still too infrequently used as a tool to shape teacher education. The first letter is from Mrs. de la Cruz, a Puerto Rican high-school bilingual-education teacher. The second letter is by Mrs. Winnebago, an American Indian teacher who teaches in a K through 8 school. The final letter is from Ms. Taylor, an African American teacher who teaches in an elementary school.

A Message from Mrs. de la Cruz

While in my teacher-preparation program, I learned mostly the mechanics of how you present material, different methodologies. When I came in as a new teacher, I wanted to follow the book-style teaching because my professors told me that this was effective. The problem is that it is very effective if you are not actually teaching any kids—if there are no kids I am sure it is very effective delivering information that way I was taught in college. However, one of the things I used to say when I started teaching was, "Here are these professors who prepared me to teach at the university. But they didn't prepare the students for me to teach." And so the students never got the same training that I did; so they weren't prepared to be taught the way that I was taught to teach. And so I found that there was a big wall between my students and me. I realized no matter what the curriculum, delivery, or classroom design, it felt like every time I came to school, I was faking it. And so I had to make the things that I taught more real and more live to the students. I had to make what I taught important and not separate from the students' lives. This is my key message. I think it can be achieved in teacher preparation if the professors perform the following tasks:

- Prepare teachers who have a sense of self, who they are, what they value, and what is important to them so that the students they teach do not see them as a fake when they enter diverse classrooms.
- Assure your future teachers are kind, open minded, have humility, and can present themselves as human beings by being calm, flexible, interested in the students and in what they teach, and are dedicated to helping students learn.
- Teach your college students to value families, parents, and grandparents along with where they come from—even if it does not match their own identities, experiences, and perspectives.
- Help your future teachers understand that they gain respect from students by not faking it and by building bonds with them. Students have to trust you to learn from you.
- Develop teachers who understand that curriculum becomes effective when whoever is teaching it and whoever is learning it buys into it. Therefore the curriculum is not "given" to students—learning goes on in all directions because we all have to buy into it.

We need more teachers dedicated to removing barriers for students. These are teachers who will not listen to the negative messages, who will respect students as they are, and who will eliminate anything that keeps them from teaching them well. These are the teachers students want to learn from.

A Letter from Ms. Winnebago

Growing up, I knew I was not White. I mean that is the biggest—that is the very first—thing that you know. Even before you start school you know that you are not White. And then you realize everything else that this means for what you will face on a daily basis from going to the store to going to school. I realized even as a child that people would not know who I am but would know that I am not White and that that would matter. You live it. You live it on a daily basis, and you know that you are not accepted for who you are. And you are always reminded of that. You can never step away from it. I think teacher educators have to make their future teachers aware of this reality for their students since it will not likely be their own. That is why they have to bring it to the forefront and not be fearful of addressing different realities. You can believe that the kids in schools are cognizant of the different realities. Frequently they bring up these issues in class. And then it spreads a little to the other kids. They have all kinds of questions about equity and injustice in our society and in their communities. I observe many teachers not being able or comfortable addressing their questions and certainly not being able to handle them in teaching and curriculum. But it is important to help students realize that society isn't equal and that they will not even have the same opportunities in school—a place they are required to attend. To me, this is the basis for working toward educational equity. We have to recognize this first. Teachers need to touch each child's life in a sense to say, "It is all right to be who you are and to be sensitive to all other people for who they are." We need to have an understanding that we should all be treated equally but have the tools to understand why we don't.

I try to get my students to believe that everybody in society—every child—is important no matter what you feel and what you think. And so by the end, hopefully, they believe that they themselves are important no matter what they think and what others think about them. It would be great if teachers believed this and valued kids who are different from them in a similar way that I teach my elementary school children. I think teacher educators can help in the following ways:

- Prepare future teachers to assure there is room for culture and identity in the curriculum and to consider that the purpose of the curriculum is to eliminate the idea that we all think the same.
- Help future teachers view the ideas, beliefs, and histories of their students as valid and important to the curriculum even when they are absent from the textbooks. Help students understand the power of textbooks to represent and privilege certain groups while ignoring and distorting others. Then use alternative sources in the curriculum so students of all groups are represented.
- Teach future teachers how to recognize curriculum differentiation based on tracking and different expectations and then to look at the curriculum more critically and to use it to alleviate such inequities.
- Prepare teachers to focus on the curriculum and why it is important that educators learn about the unjust, the unfair things that have happened in our history so that when students ask about racism and

why things are so unfair, teachers are better prepared to respond to students and to engage the curriculum with their questions rather than dismissing them, their questions, and their realities. This may enable all of us to break down some of the barriers that exist.
- [Encourage them to] be in a multicultural community themselves and assure that your future teachers experience a multicultural community—not just through placements and field experiences—but through more authentic experiences. This is essential to have a sense of who you are, to see why all groups are important, to understand the history behind our society, and to have a commitment to equity for others.

If more teachers thought of their work as essentially work for justice and equity for all people, we would have a difference in the teaching force that really mattered.

A Letter from Ms. Taylor

One of your biggest challenges in education is to help future teachers make sense of why they should teach kids who are so different from themselves, who society describes as having so much negative "baggage," and who are viewed as unworthy of what other kids deserve as a birthright. These are powerful forces that will likely influence your future teachers to produce lower expectations of students because they have such struggles and a history that is documented in their cumulative records and things like that. I would hope that your teachers don't take that information and expect less of the students and feel that they shouldn't do more with them in the way of planning and teaching. I think our struggle as teachers is to see the students we teach just as everybody sees their own kids—worthy of our best efforts. I think we struggle in the profession in not honoring diverse kids as having great potential and possibilities.

Considering that I don't see how future teachers cannot be influenced by the negative images that will certainly interfere with what they think of the children of color in their classrooms, how they plan for them, what they expect of them and their respect for them, I encourage you to prepare teachers in the following ways:

- View teaching as primarily social preparation. What I mean by that is teach them about relationships and how things relate to one another. Prepare your future teachers to teach their students about the relationship of their groups to the mainstream. Teach them that they will encounter the mainstream of society and how to do so and maintain their own sense of self, identity, and history.

- Determine strategies to prepare teachers to become members of the community in which they teach but likely will not live. This is essential to being able to make the school curriculum relevant and not based on stereotypes about the kids and their communities. It is further essential to genuinely understand the students' realities, interests, and sense of identity. And finally it is a prerequisite to being able to create a learning environment in which students can question things, express their own views, and believe that they know something valued within the classroom.
- Be independent thinkers based on reality, information, and the desire to learn more about who they teach and their assets and abilities. Help them understand where their perspectives on the diverse groups they will teach come from. Are they from mainstream media and personal experience, for example? Expose your future teachers to more indigenous resources that do not present negative, stereotypical images of those they will teach. And prepare future teachers to want their own students to be independent thinkers who can think through an issue, have an opinion about it, and want to know more about it rather than being compliant and passive.
- Prepare your future teachers to listen to the realities of the students they teach without judgment and without developing pity, seeing them as victims, or taking on a missionary approach to teaching. Instead teach them to create culturally centered curricula so student voices are included. Prepare them to use this to help their own students understand their realities and how they relate to others. Regularly say to students, "Everybody give me a story" as a means to move beyond superficial knowledge of who the students are, what their realities are, and how you can connect the curriculum. I started listening to other people's stories and that helped me to begin to understand my participation historically in this country and around the world.
- Start at an early age to expose children to their group's history to assist in making the curriculum relevant to developing students' sense of self and history. Give them ideas to question, to start them thinking and wondering where am I and how do I fit.

Schools have to change. Teachers have to change. Someone asked a counterfeiter, "How do you know counterfeit money?" He said," I study good money." All teachers need to study good teaching and to use it to address inequities.

Reframing Teacher Education from Practice to Theory

The above letters are extremely powerful. The voices, perspectives, and insights these teachers offer to teacher educators contribute to how the nation's teachers should be prepared now and in the future, because the racial, ethnic, and linguistic mismatch is omnipresent. Lincoln and Denzin (1994) express the significance of these voices when they state, "A politics of liberation must always begin with the perspectives, desires, and dreams of those individuals and groups who have been oppressed by the larger ideological, economic, and political forces of a society, or a historical moment . . . The center shifts and moves as new previously oppressed or silenced voices enter the discourse" (p. 575). As educational equity continues to be debated and sought, voices from the margins increase in their value. I argue that these three teachers of color represent three voices from the margin considering the profile of the teaching force in the United States. The work presented here is one attempt to unsilence voices that have a limited presence in education discourse by the sheer lack of their quantitative presence and by their marginalization as teachers of color.

The letters of these teachers stand alone. To analyze them seems unnecessary and a violation of their integrity and power to speak for themselves. So rather than do so, I will explore what I hear from these organic intellectuals to move the preparation of teaching into a discourse of political liberation through multiple discourses. Essentially, I will explore how these "organic intellectuals" (Gramsci in McLaren, 2003, p. 48) can help the profession by speaking from the inside out. That is, how do their voices inform practice to theory rather than the oft-held view that teachers, and in particular teachers of color, are practitioners who are not theory builders and who do not connect their work to larger sociopolitical discussions. Their voices are essential because they are shaped by their racial, ethnic, and linguistic positionalities and identities. Banks (2006) reminds us that our identities are composed of our group's values and behavioral styles, perspectives, worldviews, frames of reference, identification, cultural cognitiveness, language and dialects, and nonverbal communications. He wrote, "These components are useful for interpreting the behavior of students and teachers" (p. 74). That is exactly why these letters are so important. They assure a place and a voice in teacher preparation for these diverse voices, for the many teachers like them who preceded them, and for the few who will follow. McLaren (2003) declared that "voice is not a reflection of the world as much as it is a constitutive force that both mediates and shapes reality within historically constructed practices and relationships shaped by the rule of capital" (p. 245). Thus, their narratives (presented through the above letters) are not simply their words but are certainly about their realities in a sociopolitical context like that in the United States that produces the racial, ethnic, and linguistic mismatch that is the focus of this chapter. Voice, then, is the suggested means by which these teachers and other teachers of color have to make themselves heard and to define themselves as active participants in the world (McLaren, 2003) even when they experience increasing physical invisibility.

Three key themes to reframe theory from practice emerged from the letters of these teachers (see Table 9.1). Each of these separately is important in challenging the current paradigms in teacher preparation, but together they hold the potential to have a significant impact not only on the work of teacher educators but also on their role beyond their campuses. Teacher educators have worked extensively and with sincere dedication to prepare teachers for diverse classrooms and schools. Judith Lanier (2007) explained in the new Foreword to the *Holmes Trilogy* that, after a generation of reforming teacher education, "the agenda is important and remains unfinished . . . the original *tomorrows* are now yesterdays, and the future for many students is not promising" (p. xxv). This certainty serves as the backdrop to hear how these teachers' voices help readers reframe the next phase of incomplete tomorrows.

Traditionally, in the old frame, "diversity of others," teacher educators have worked extensively to prepare future teachers for diverse schools and classrooms to teach "the other." These efforts have taken various practical approaches emanating from various ideologies about what diversity means. Much of that work has focused teacher preparation on teaching about various minority groups. This work often appears to parallel the curriculum of many schools themselves where the school year is essentially a series of distortions about various groups under the umbrella of diversity. For example, the school year begins with teaching distortions about American Indians, followed by African Americans, Latinos, and other groups. The university teacher-education curriculum too closely mirrors this pattern in that some professors employ this serialized "othering" of groups under the pretense of teaching students about the various groups they will encounter in their classrooms. The result in both cases, in schools and in teacher education, is too frequently teaching racism, distortions, and miseducation. It

Table 9.1. Practice-to-theory implications for teacher education

Old frame	*Equity reframe*
Diversity of others	Shared fate of all
Individual achievement	Opportunity for all
Problem-focused diversity	Solutions and community benefits for all

relies on a human-relations approach to diversity that "prepares teachers to honor diverse student backgrounds and to promote harmony among students" (Banks & Tate, 1995, p. 147).

The teachers of this study seem to suggest reframing the teaching of diversity from notions of harmony and narrow distortions of group characteristics of "other" minority groups to focusing instead on how the fates of all groups are shared. Instead of running through what is referred to as the "parade of the cultures," the focus would be reframed on structural flaws in our society that lead to various disparities and inequities and the negative impact that results for everyone. The focus shifts to the struggles in our society, systems that perpetuate them (including education), and how disparities can be analyzed systematically. The result of this analysis should not be hand wrenching or, in some cases, tears of sadness and pity. The result should be preparing teachers with a sense of agency and who are solutions focused.

Individual achievement symbolizes the quintessential core of education. In fact, there is little room in education dialogue to consider school achievement as anything other than individual success or failure regardless of any other communal identities that might be important to growth and development. Even after decades of incorporating various forms of cooperative work into teaching practices, achievement is still viewed as ultimately individual student performance. Further, achievement is viewed as benefiting the student only and does not extend to anyone else, whether that be to the community or to the larger society. McLaren (2003) describes this educational romance of individual achievement as one of the biggest myths in education that leads us to conclude that achievement or underachievement is individual or personal.

I hear the teachers in these letters encouraging a reframing of individual achievement in order to interrupt its dominance as an ideology about the foundation of schools and American education. They conversely speak of education in a different paradigm, where the opportunity to learn and to achieve should be accessible to everyone and of benefit to everyone—individually and as a community. This idea is already fundamental in some communities of color and was well articulated by the teachers of this study. They were clear that groups, communities, and societies benefit when racial, ethnic, and linguistic inequities are taken up seriously by the work of teachers. Therefore teachers not only engage in producing high achievement for students as individuals but also see access to high achievement for all students as an equity issue for the larger society.

After generations of purposeful attention to preparing teachers for diverse classrooms and schools, the ideology of diversity as a problem still afflicts much of teacher education. Problem-focused diversity describes teacher-education contexts where diversity is viewed as a challenge for teaching. This is sometimes couched under the paradigm of celebrating diversity, but the message is clear that diversity poses problems and will require some form of adapting pedagogy and curriculum. In teacher-preparation programs where teachers are implicitly taught that diversity is a challenge, teachers are prepared to have certain fears of the diverse groups they will teach as well as uncertainty about their ability to teach them. Teachers completing programs grounded in this paradigm may question their preparedness and efficacy in teaching diverse groups and may enter classrooms with much trepidation and may fault the groups for being a problem in society.

The three teachers in this study ask that we reframe diversity away from a problem for teachers and their work and instead toward a perspective on education as a form of social action and agency in the classroom and in the larger community. Teacher preparation can prepare teachers to move beyond adumbrating the problems and obstacles of diversity. They can prepare teachers who view their work as benefiting the individual student as well as the broader community. This has the potential to anchor teachers and to give them a stake in teaching beyond simply academic subjects. This stance is essential to equity in education.

Reframing teacher preparation is a journey that began some time ago. It is an incomplete journey, and it is a journey facing exigency and harsh critique. The voices of the teachers represented here offer a powerful juncture for teacher educators to sharpen, redirect, and deepen the ideas essential to a true reframing of teacher education and its relevance to educational equity. Their letters issue a call for us to hear them, to listen to them, and to act with them to use our sphere of influence and knowledge to march on for equity. We do not have to abandon this ideal.

References

AACTE. (2004). *Culture, language, and student achievement: Recruiting and preparing teachers for diverse students.* Washington, DC: Author.

Au, K., & Jordan, C. (1981). Teaching reading to Hawaiian children: Finding a culturally appropriate solution. In H. Trueba, G. Guthrie, & K. Au (Eds.), *Culture and the bilingual classroom: Studies in classroom ethnography* (pp. 69–86). Rowley, MA: Newbury House.

Baldwin, J. (1985). *The price of the ticket: Collected non-fiction, 1948–1985.* New York: St. Martin's. (Reprinted from *A talk to teachers*, The Saturday Review, December 21, 1963)

Banks, J. A. (2006). *Cultural diversity and education: Foundations, curriculum and teaching.* Boston: Pearson Education.

Banks, J. A., & Tate, W. F. (1995). Multicultural education through the lens of the multicultural education research literature. In J. A. Banks & C. A. M. Banks (Eds.), *Handbook of research on multicultural education* (pp. 145–166). New York: Macmillan Publishing.

Clewell, B. C., Puma, M. E., & McKay, S. (2001). *Does it matter if my teacher looks like me? The impact of teacher race and ethnicity on student academic achievement*. Washington, DC: The Urban Institute.

Foster, M. (1994). Effective Black teachers: A literature review. In E. R. Hollins, J. E. King, & W. C. Wyman (Eds.), *Teaching diverse populations: Formulating a knowledge base* (pp. 225–241). Albany, NY: State University of New York Press.

Gay, G. (2000). *Culturally responsive teaching*. New York: Teachers College Press.

Gordon, B. M. (1995). Toward emancipation in citizenship education: The case of African American cultural knowledge. *Theory and Research in Social Education, 12*(4), 1–23.

Gramsci, A. (1971). *Selection from the prison notebooks*. London: Lawrence & Wishart.

Hess, F. M., & Leal, D. L. (1997) Minority teachers, minority students, and college matriculation: A new look at the role-modeling hypothesis. *Policy Studies Journal, 25*, 235–248.

Lanier, J. (2007). Foreword. In *The Holmes Partnership trilogy: Tomorrow's teachers, tomorrow's schools, tomorrow's schools of education* (pp. xi–xxv). New York: Peter Lang.

Irvine, J. J., & Armento, B. J. (2001). *Culturally responsive teaching*. Boston: McGraw Hill.

Jordan, C. (1985). Translating culture: From ethnographic information to educational program. *Anthropology and Education Quarterly, 16*, 105–123.

Keleher, T., Pina, L. D., & Fata, M. G. (1999). *Creating crisis: How California's teaching policies aggravate racial inequalities in public schools*. Oakland, CA: Applied Research Center.

Ladson-Billings, G. (1994). *The dreamkeepers: Successful teachers of African American children*. San Francisco: Jossey-Bass Publishers.

Lincoln, Y. S., & Denzin, N. K. (1994). The fifth moment. In Y. S. Lincoln & N. K. Denzin (Eds.), *The handbook of qualitative research* (pp. 575–586). Thousand Oaks, CA: Sage.

McLaren, P. (2003). *Life in schools*. Boston: Pearson Education.

Myrdal, G. (1944). *An American dilemma: The Negro problem and modern democracy*. New York: Harper.

Meier, K. J., & Stewart, J. (1991). *The politics of Hispanic education: Un paso pa'lante y dos pa'tras*. Albany, NY: State University of New York Press.

National Center for Educational Statistics (NCES). (2007). *The condition of education: Contexts of elementary and secondary education*. Washington, DC: U.S. Department of Education.

Pallas, A. M., Natriello, G., & McDill, E. L. (1989). The changing nature of the disadvantaged population: Current dimensions and future trends. *Educational Researcher, 18*(5), 16–22.

Patton, M. Q. (2002). *Qualitative research and evaluation methods.* Thousand Oaks, CA: Sage.

Siddle-Walker, V. (2001). African American teaching in the South: 1940–1960. *American Education Research Journal, 38*(4), 751–779.

Valenzuela, A. (1999). *Subtractive Schooling: US Mexican Youth and the Politics of Caring.* New York: SUNY.

Vogt, L., Jordan, C., & Tharp, R. (1987). Explaining school failure, producing school success: Two cases. *Anthropology and Education Quarterly, 18*(4), 276–286.

Afterword

Sonia Nieto, University of Massachusetts

In these times of rapidly changing demographics, rigid accountability structures, and widespread standardization, preparing teachers for diverse classrooms is more than just a challenge; it is a duty. As the authors of the chapters in *Culture, Curriculum, and Identity in Education* have made clear, reasons beyond the "demographic imperative" (Banks, 2006; Cross, Chapter 9 in this volume) make preparing teachers for today's classrooms an urgent necessity. Given our responsibility to the next generation, it is also an opportunity. Students today attend schools that are far more complex and bureaucratic with more stringent demands and greater expectations than ever before. At the same time, the so-called achievement gap (what I have elsewhere called the "resource gap" or the "caring gap"; Nieto, 2006) is improving only negligibly, resulting in more students of color dropping out of school and unprepared for life and work.

The authors of this book take on the challenge of preparing teachers for diverse classrooms in various ways. They consider the important connections inherent in culture, curriculum, and identity in education. Some address the sociopolitical context of education; others explore the obligation of the university, and specifically the teacher education program, to prepare future and practicing teachers with the skills and attitudes to serve all their students equitably; still others delve into the meaning of spirituality for the preparation of teachers. What all these chapters have in common, in spite of these differences, is their attention to the reality of the growing diversity in our nation's schools—a reality that has, up until now, been only partially addressed by the field of education. Although many more schools and colleges of education, as well as district professional-development programs, are taking seriously the question of how teachers might be better prepared to serve students of diverse backgrounds, in general the results have not been terribly promising in terms of either teacher knowledge or student learning. Some clues as to why this is the case are the focus of the remainder of this Afterword. Specifically, I want to draw a number of lessons to be learned from the book's authors for theory and practice, both for teacher educators and for classroom teachers.

Lessons to Be Learned

Many suggestions for improving teacher education and P through 12 education with a focus on diversity and equity are embedded within the various chapters in this book. These suggestions point to needed changes in teacher preparation and professional development as well as in teachers' values and practices in their respective classrooms. In particular, in what follows I explore three interrelated themes that are paramount in the book: *the significance of context*; *the political nature of education*; *and why relationships are at the center of teaching and learning*. In the end, these suggestions point to access and equity as the key components of a social-justice perspective in teaching.

Context Matters

Where, *how*, and *under what conditions* education takes place are crucial elements to consider in preparing teachers for equity and diversity and for teachers to understand in the P through 12 classroom. That is, education does not happen in a vacuum but rather in a particular place and time and under particular conditions. Take the issue of dropouts, for example. Currently, about half of all African American, Latino, and American Indian students drop out of school before graduation (Orfield, 2004). While most educators acknowledge the crisis in graduation rates, it is most often viewed as a problem residing in the students themselves—that is, their lack of effort, motivation, or intelligence. Yet the social, political, and economic context explains dropping out as much, or even more so, than individual behaviors or attitudes of students. Moreover, although student characteristics (poverty, parental education, family structure, English-language ability, and others) certainly play a role in student learning, they alone are not the reason for dropping out of school. Conditions in schools such as the nature of the curriculum, the material resources available to students, the professional development opportunities for teachers and administrators, the level of school funding, and class size, among many other institutional policies and practices, can fairly reliably predict student outcomes.

The fact that "context matters" is evident in many of the chapters in this book, and especially in the chapters by H. Richard Milner (Chapter 2), Jason Irizarry and John Raible (Chapter 4), and Carl E. James (Chapter 5). Although Irizarry and Raible's chapter focus is ELL students, the message they send applies to all contexts. For example, rather than simply present the demographics of ELLs—as if these students had just dropped from the sky—they explore the history of U.S. colonization and imperialism, as well as globalization and the experiences of Latinos and Asians, currently the largest immigrant groups in our nation, within that history. Without this context, issues such as the frenzy over immigration policy or the debate over bilingual education would be difficult to understand. In the same way, this kind of context provides a foundation for understanding the experiences of, for instance, a recent immigrant Mexican student that Iddings and Rose (in Chapter 3) introduce to readers of this volume. Thus, according to the analysis of Irizarry and Raible, without an explicit knowledge of the sociopolitical context of education, teachers and teacher educators alike will be unable to prepare their students to face the challenges of life in the twenty-first century.

Education Is Political

Closely related to the issue of context is the political nature of education. Many years ago, Paulo Freire reminded us that education is always political (Freire, 1970). Keeping in mind the political nature of education is a message shared by many of the authors in this volume, and it is a good reminder for all of those who prepare teachers and for teachers themselves to consider. Put another way, power relationships play a crucial role in all aspects of education. This means that every educational decision we make, whether it concerns the choice of a reading program or a plan to promote community outreach, whether it is about which history textbook to buy or who will be studied as "community heroes," reflects the decision maker's view of the world. Decision makers can be found, of course, at various levels: the classroom; the school; the district, city, state, and federal governments; as well as in the academy—including in teacher education programs. And while it is true that institutional and federal structures wield greater power than individual teachers or administrators, individuals can also make a difference because they make numerous decisions every day that influence the quality of life for students in school. For example, the kind of research in which teacher educators choose to engage says something about their ideological position and the connections they see among themselves, the teachers they prepare, and the communities to which they send these teachers, a point addressed in Michael Dantley's chapter (see Chapter 6).

How can teacher-education programs take into account the political nature of education? Although teacher preparation has traditionally viewed education as apolitical, more programs are recognizing—and preparing future teachers to recognize—that education is a fundamentally political endeavor. In such programs, prospective teachers are encouraged to think about equity and social justice as fundamental to education. They are expected to look at their own practice, and the policies and practices of their schools and beyond, not as finished and static but as changeable and dynamic. One way in which teacher-education programs can acknowledge the political nature of education is by providing courses that include a historical and sociological analysis of education. Education courses themselves have changed in the past two decades to encompass a broader view of content. For example, texts such as Ron Takaki's *A Different Mirror: A History of Multicultural America* (1993) and Joel Spring's *Deculturalization and the Struggle for Equality: A Brief History of the Education of Dominated Cultures in the United States* (2006) have become standard texts in many teacher education classes.

The message that education is political is evident in Beverly Cross's chapter, particularly in the letters written by teachers of color to teacher educators. The advice they provide in their letters is invaluable, and I wish all teacher educators had the opportunity to read them. Witness, for example, Ms. Winnebago's statement on the political nature of knowledge: "Help students understand the power of textbooks to represent and privilege certain groups while ignoring and distorting others." If all future teachers were to learn this lesson, we would have come a long way in promoting knowledge as multifaceted and complex. The teachers' letters also address the attitudes, values, and sensibilities they believe make teachers effective with their students of color. For instance, Mrs. de la Cruz writes, "Prepare teachers who have a sense of self, who they are, what they value, and what is important to them so that the students they teach do not see them as a fake when they enter diverse classrooms."

Relationships Are at the Heart of Teaching

That relationships are at the heart of teaching is a message evident in many of the chapters in *Culture, Curriculum, and Identity in Education.* One way in which this point is apparent is in the attention given to student identities and the assertion that teachers, in order to be effective, need to be both aware of and affirming of student identities. Stephen Hancock (in Chapter 7) addresses the connection between spirituality and morality—two issues often left out of educational conversations. Also, as Eric Toshalis writes, "Urban teaching is identity work." Nothing could be truer. Unfortunately, in some teacher education programs, future teachers are still admonished to be "color-blind" and simply ignore student differences. Although being color-blind can be a virtue when it means that people are not judged negatively by the color of their skin, their social class, or the language they speak, it can also be an obstacle to forming relationships with students when it is used as a way of denying or ignoring differences as if they were negative to begin with.

Developing strong and meaningful relationships with students means, first of all, recognizing who students are and, secondly, accepting this reality. Only when this happens can the work of teaching and learning proceed. Language, for instance, which is too readily "corrected" before teachers have accepted and affirmed students' identities, is a case in point. This does not mean that students need not learn the "standard" variety of the language—in fact, they *must* learn it if they are to make their way in the world—but rather that before they can do so, they need to have their way of communicating accepted as valid. Paulo Freire (1998), in a letter written to teachers in Brazil, could just as easily have been addressing teachers working in poor urban or rural areas in the United States:

> When inexperienced middle-class teachers take teaching positions in peripheral areas of the city, class-specific tastes, values, language, discourse, syntax, semantics, everything about the students may seem contradictory to the point of being shocking and frightening. It is necessary, however, that teachers understand that the students' syntax; their manners, tastes, and ways of addressing teachers and colleagues; and the rules governing their fights and playing among themselves are all part of their *cultural identity*, which never lacks an element of class. All that has to be accepted. Only as learners recognize themselves democratically and see that their right to say "I be" is respected will they become able to learn the dominant grammatical reasons why they should say "I am." (p. 49)

Another issue addressed in several of the chapters is the significance of the identity of teachers themselves (see Milner's Chapter 2, for instance). Overall, the authors make three points clear: First, teachers' identities matter. It is no longer acceptable (if it ever was) for teachers to assume that either their identities or their students' identities are "generic," as if their race, ethnicity, social class, language, gender, sexual orientation, or other markers of identity did not influence their ideas and practices. Second, these authors emphasize that *all* teachers can be effective with students of color. Although identity matters, it is not the only thing that matters. This is the case, for instance, in Chapter 5 by Carl James in which he recounts the experiences of Conrad and Kendra, two African Canadian young people for whom context was a pivotal factor in moving them from a working-class "minority" community outside of Toronto to the university they were currently attending. The third point concerning identity made by the authors in this collection is that schools must be places where students see and interact with teachers and other staff members who look like them and share similar experiences. This means that the recruitment of teachers of color is crucial. Until the dearth of teachers of color in our nation is faced and remedied, equity will continue to be elusive.

Conclusion: Equity and Access Are Key to a Social-Justice Perspective

It is obvious that *Culture, Curriculum, and Identity in Education* addresses issues with which all P through 12 schools and colleges of education must grapple. Rather than a simple addition of a course on multicultural education or a mission statement that includes the words "social justice," *Culture, Curriculum, and Identity in Education* implies an overhaul of colleges and schools of education. This means reexamining goals, rethinking strategies for recruiting a more diverse student body and faculty, reforming the teacher-education curriculum, developing more meaningful field placements, and strengthening relationships with the communities to which we send future teachers. It means "placing equity front and center" in P through 12 schools and teacher education, an issue I have explored in more depth elsewhere (Nieto, 2000).

Providing all young people with an equitable education and with access to the benefits of a full and meaningful life is what social justice needs to be about. In order for these things to happen, we have to begin with the attitudes, values, skills, and competencies teachers bring with them into the classrooms. They deserve a preparation based on respect and trust. They deserve an education that prepares them to meet the challenge to educate *all* students well. In the final analysis, not only will they be empowered through such an education, but so will their students. And that, of course, is the bottom line.

References

Banks, J. A. (2006). *Cultural diversity and education: Foundations, curriculum, and teaching*. Boston: Allyn & Bacon.

Freire, P. (1970). *Pedagogy of the oppressed*. New York: Seabury Press.

Freire, P. (1998). *Teachers as cultural workers: Letters to those who dare teach*. Boulder, CO: Westview Press.

Nieto, S. (2000). Placing equity front and center: Some thoughts on transforming teacher education for a new century. *Journal of Teacher Education, 51*(3), 180–187.

Nieto, S. (2006). Solidarity, courage, and heart: What teacher educators can learn from a new generation of teachers. *Intercultural Education, 17*(5), 457–473.

Orfield, G. (Ed.) (2004). *Dropouts in America: Confronting the graduation rate crisis.* Cambridge, MA: Harvard Education Press.

Spring, J. (2006). *Deculturalization and the struggle for equality: A brief history of the education of dominated cultures in the United States* (5th ed.). New York: McGraw-Hill.

Takaki, R. (1993). *A different mirror: A history of multicultural America.* Boston: Little, Brown, & Co.

Contributors

Beverly Cross is the Lillian and Morrie Moss Chair of Excellence in Urban Education at the University of Memphis. Her research interests are in urban education, teacher education, and curriculum development.

Michael E. Dantley is associate provost and associate vice president for academic affairs and professor of educational leadership at Miami University, Oxford, Ohio. His research focuses on leadership, spirituality, and social justice.

Ana Christina DaSilva Iddings is an assistant professor of language, literacy, and culture in the Department of Teaching and Teacher Education at the University of Arizona. Her research interests are language learning, immigration and equity in education, and the preparation and professional development of teachers to work with English language learners.

Jason G. Irizarry is an assistant professor of multicultural education in the Department of Curriculum and Instruction in the Neag School of Education at the University of Connecticut. A central focus of his work involves promoting the academic achievement of Latino/a and African American youths in urban schools by addressing issues associated with teacher education.

Jacqueline Jordan Irvine is the Charles Howard Candler Professor of Urban Education Emerita in the Division of Educational Studies at Emory University. She also holds visiting appointments at the University of Maryland and Howard University. Dr. Irvine's specialization is in multicultural education and urban teacher education, particularly the education of African Americans.

Stephen D. Hancock is an assistant professor of education in the Department of Reading and Elementary Education at the University of North Carolina–Charlotte. His research interests are effective teaching in urban schools, teacher efficacy, multicultural issues in education, and the intellectual, spiritual, and social development of urban students and teachers.

Carl E. James is a professor in the faculty of education at York University, Toronto, Canada. Currently, he is the director of York Center for Education and Community. In his research, he explores issues of equity related to race, ethnicity, class, and gender; youth and sports; access to postsecondary education; and multicultural/diversity (including affirmative-action) programs and practices.

Sonia Nieto is professor emerita of language, literacy, and culture at the University of Massachusetts–Amherst. She has taught students at all levels from elementary through graduate school, and she continues to speak and write on multicultural education, teacher education, the education of Latinos/as, and other culturally and linguistically diverse student populations.

John W. Raible is assistant professor of diversity and curriculum studies in the Department of Teaching, Learning, and Teacher Education at the University of Nebraska–Lincoln. His research focuses on multicultural education and family diversity, multiple identities in transracial adoptive families, and other interracial contexts.

Brian C. Rose is a doctoral student in the Department of Teaching and Learning in the Language, Literacy, and Culture Program at Peabody College, Vanderbilt University. His research interests include second-language acquisition, second-language pedagogy, and the preparation and professional development of teachers who instruct ethnically and linguistically diverse students.

Eric Toshalis is assistant professor of secondary education at California State University Channel Islands. His research investigates overlapping factors that influence teacher-student relationships in urban public schools, namely, disciplinary interactions, developmental trajectories, processes of identity formation, pedagogical strategies, and differences in culture, ethnicity, gender, language, race, sexuality, and socioeconomic class.